AMUR TIGER
David Prynn

The author has studied Amur tigers and leopards over many years. In the course of doing so he learnt Russian and visited the Russian tiger country. He describes the tiger and its ecology, as well as the history of its study by Russians, and prospects for its future. Seven chapters are stories about tigers translated from the Russian literature.

The book will appeal to the non-specialist reader with an interest in tigers, while for the specialist there is extra information in Notes and Appendices.

Дэйвид Прынн

АМУРСКИЙ ТИГР

David Prynn

AMUR TIGER

Translations by G.H.Harper

Russian Nature Press
2004

Published in 2004
by
Russian Nature Press
19/3 Warriston Road
Edinburgh EH7 4HN
United Kingdom

ISBN 0 9532990 3 1

A CIP catalogue record for this book
is available from the British Library

Designed by RNP and
set in 11/12pt & 11/12.5pt Garamond

Printed and bound by
Short Run Press Ltd
Exeter, Devon

CONTENTS

Maps

PUBLISHER'S NOTE

RNP is pleased to add *Amur Tiger* to its list. It is the first RNP book to cover a single species, and also the first to deal in any depth with the Russian Far East. Other mammals of the former Soviet Union are described in Vladimir and Yuri Smirin's *Animals in Nature* - which incidentally has a chapter on the gorals of the Sikhote-Alin and Lazovski zapovedniks (strict nature reserves). The zapovedniks in general are covered in Felix Shtilmark's *History of the Russian Zapovedniks 1895-1995*. (See pp.210ff for details about RNP books.)

The most important chapters have been written by David Prynn - an expert on the Amur tiger - and his name rightly appears on the title page as author. For practical reasons certain chapters have been written by the translator, since some material was available only in Russian and only in Moscow.

Amur Tiger can be read in a number of ways. The stories are the even-numbered chapters. They are quite independent and they require no specialist knowledge, although some is provided in the Notes at the back of the book. The remaining chapters give information about the tiger, its habitat and ecology, the Russian Far East, and the tiger's prospects of survival. These chapters are also largely independent. So you can tackle chapters in any order you like. Apart from the technical information and references given in the Notes - which can be ignored by the non-specialist - other information is summarised in the four Appendices. In case you get lost, the Index will help you navigate.

Publication has been timed to coincide with the exhibition 'Cats' - organised by Dr Andrew Kitchener at the Royal Museum, Edinburgh - from February to May 2004.

RNP is a not-for-profit company with the charitable aims of promoting ecology and conservation in the former Soviet Union. It depends on sponsorship to remain afloat, and we therefore wish to record our gratitude to the late **Mrs Hetty Harper** for sponsoring *One Season in the Taiga*, **Dr and Mrs David Boddington** for *Animals in Nature*, and an **anonymous sponsor** for *History of the Russian Zapovedniks*. We are also indebted to **Donald Elliot** (Elliot & Company WS, 8 Charlotte Street, Perth PH1 5LL), who donated his services when RNP was incorporated as a Company Limited by Guarantee as a first step to acquiring formal charitable status. Sponsorship usually takes the form of paying the printing bill for a book or giving equivalent assistance.

Our thanks go to Becky Govier (Royal Botanic Garden Edinburgh) for assistance in seeking sponsorship, Nikolai Formozov and Ekaterina Pavlova for permission to publish a translation of their obituary of Evgeni Matyushkin, Andrew Kitchener (National Museums of Scotland) for reviewing some chapters, Richard Lines, who revised the whole book, Yuri Smirin for stimulating comments about the Amur tiger's prospects as seen from a Russian's point of view, Evgeni Smirnov (Sikhote-Alin Zapovednik) for permission to translate his article in chapter 14, Pavel Tomkovich (Zoology Museum, Moscow) for providing the obituary of Evgeni Matyushkin, and Mike Wilson (Edward Grey Institute, Oxford) for lending some books about the Russian Far East.

RNP

AUTHOR'S NOTE

Why is the Amur tiger special and why do we need a book about it? All tigers are endangered, but the Amur tiger is one of the rarest, despite the great efforts being made to save it. It is different from other tigers - a massive long-furred animal adapted to snowy winters at the northern limit of the species' range. This impressive big cat, more than once threatened with extinction in the twentieth century, is significant as a flagship for the protection of the beautiful mixed forests - of great biological diversity and containing many rare species - in the southern Russian Far East and Northeast China. The Amur tiger - called 'Amba', the great sovereign or master of the forests by the native people - is also important as a cultural icon and a magnificent symbol of the heritage of the indigenous people, Russians, Chinese and Koreans, giving special significance to the region.

Although the Amur tiger has been extensively studied in the wild by Russian scientists since the 1940s, very little of their published research appeared in English - that is, until the work in the 1990s of the US Hornocker Wildlife Research Institute, now merged with the Wildlife Conservation Society.

For many years I have been developing a deep interest in the little-known regions of the southern Russian Far East and Northeast China, and their unique wildlife, especially the Amur tiger. This interest was enhanced by a number of study tours to most of the reserves inhabited by the great cats. With the collection of evidence from museums, libraries and zoological collections in Russia and elsewhere I came to appreciate that the Amur tiger was in grave danger, especially on account of poaching for its skin and body parts used in traditional oriental medicine. I also realised that little was known in the West of its status and interactions with people. So it seemed logical to develop this material in a study of the tiger and its importance to conservation.

I owe a great personal debt to those who have helped me in my research over the years. I should like to thank and acknowledge the work of the intrepid Russian scientists who have spent their lives in the field, particularly Evgeny Matyushkin, Dimitri Pikunov, Evgeny Smirnov, Galina Salkina and the late G.F. Bromley. The effective field work and conservation efforts of expert US biologists Dale Miquelle, John Goodrich and Linda Kerley have built on the achievements of their Russian colleagues, and their writings and incisive comments have been an invaluable source of information and insight for me into the life of the great striped cat. Complementing these remarkable efforts in the wild have been the contributions of the conservation-body coordinators, who are working so hard to save the Amur tiger and the other wildlife of the region. I have deeply appreciated the kind assistance and inspirational leadership of Sarah Christie (21st Century Tiger), the courage and dedication of Irina Melenina (Phoenix Fund of Vladivostok), the unrivalled contribution to wild cat conservation of Peter Jackson (former Chairman of the Cat Specialist Group of the International Union for the Conservation of Nature) and the tireless efforts of Sharon Miller (AMUR Trust).

In conclusion I should especially like to express my gratitude to Arthur Goldwater and to Ann Stannard, without whose assistance the writing of the book would not have been possible. My hope is that this modest study will make some contribution to the efforts to save this magnificent animal and the wonderful diversity of life that constitutes the Russian Far East forests.

David Prynn

TRANSLATOR'S NOTE

The role of translator may seem a humble 'back-room' affair, but the work for this book has been anything but hum-drum. Just trawling through Vladimir Arsenev's books for suitable material has been hugely enjoyable, and even exciting. While his story of Dersu Uzala is well known and has been translated into English, the accounts of his other expeditions appear not to have been. So I have read them all, following his every step while exploring the Sikhote-Alin Mountains, suffering with him in his calamities, and have been fascinated and moved by his accounts of what was then virtually unknown tiger territory. Quite possibly I was the only non-Russian to have read these books for many years, and felt I was sharing in his adventures and discoveries.

So I first want to thank Oleg Kosterin, Pavel Tomkovich, Irina Utekhina and other Russian friends who found for me the Arsenev editions and many other Russian works describing that country's wonderful natural history. WWF Russia Programme Office kindly provided copies of Evgeni Matyushkin's *The Amur Tiger in Russia.* Indispensable for collecting material was a week in Moscow in late January 2003, when I was privileged to stay with Feliks and Nadezhda Shtil'mark, and make use of Feliks' impressive library. He has also answered a number of technical queries that arose during translation. Dr David Boddington explained why the skin should be rubbed with snow in certain circumstances, and finally Dr Hongxing Zhang (National Museums of Scotland) helped investigate Chinese dedications to the tiger found by Arsenev in the Sikhote-Alin Mountains.

Now, alas, we come down to earth and broach the subject of transliteration - since, sadly, it is impractical to use here the attractive Russian cyrillic alphabet for Russian words. This alphabet has several more letters than the English one, and so no system of transliteration can be perfect if it uses just English letters. Most consonants can be pronounced as in English, except that '*zh*' sounds like the second 'g' in French 'garage', and '*kh*' sounds like 'ch' in German or Scottish 'loch'. As for vowels, '*e*' is like 'ye' in 'yes', '*i*' is like 'ee' in 'keen'; '*y*' before a vowel is like 'y' in 'yes'; '*y*' on its own is a bit like 'i' in 'bit'. The diphthong '*ay*' is pronounced 'I' or 'aye'; thus '*tugay*' sounds like '2-guy'. One sound is always stressed - '*ë*', like 'yo' in 'yonder'. The transliteration system is slightly simplified in the main text, but is given more fully in Appendix 1, References, and Notes (where an apostrophe indicates the hard or soft sign). An underlined vowel shows that the syllable is stressed.

Geoffrey Harper

Map 1. *The former Soviet Union*

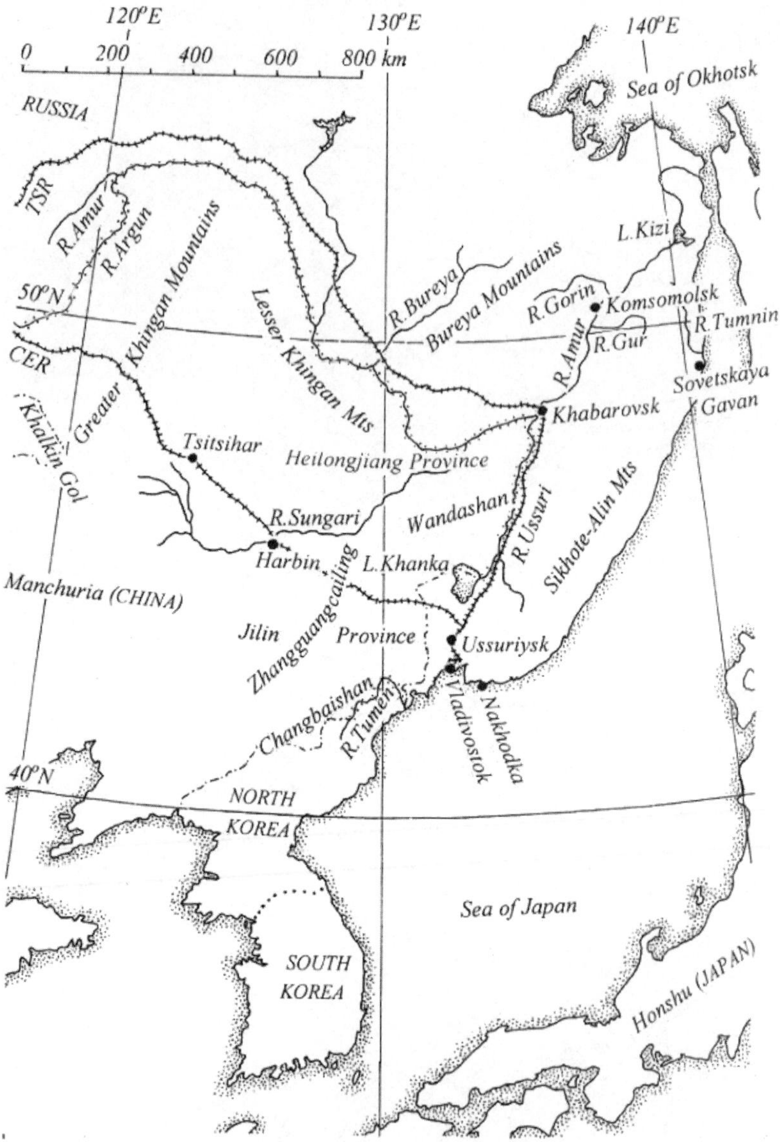

Map 2. The southern Russian Far East, north-east China, and Korea

Chapter 1
INTRODUCTION

This book is about the Amur, Ussuri or so-called Siberian tiger. The Amur and Ussuri rivers are within the tiger's historical range, so are acceptable names; 'Siberian' is not recommended, because the tiger has probably never occurred in Siberia as normally understood in Russia. The book does not aim to describe tigers in general, since there are several excellent introductions to the species as a whole.[1] To set the Amur tiger in context, however, information about all the main kinds of tiger is summarised in Appendix 2.

The book is ambitious. We have tried to describe, however briefly, everything worth saying about the Amur tiger in the wild. Where it *is* only a brief mention, there will often be a work cited in the References to enable you to find out more information. As is to be expected, much space is given to the tiger's biology (chapters 9 and 11). Because conservation is essentially based on ecological knowledge, the tiger's prey, competitors and habitat are described in some detail. These depend on vegetation, climate and geography, which are covered briefly in chapter 3. So, by the end of the book you will appreciate why the tigers depend on the oak and pine trees.

Even more than the natural environment, it is humankind that will now be the deciding factor in whether the Amur tiger survives in the wild. The human-tiger inter-relationship has been approached from several angles. Much has been written about the tiger in Russian, and the even-numbered chapters give a flavour of the tiger as it appears in Russian literature. The historical relationship between tiger and people is touched on in chapter 3, and in more detail in chapters 5 and 12. The scientific study of tigers in the field is described in chapters 7 and 14. Finally, present-day interactions between tigers and the human population are the subject of chapter 15. The many non-governmental organisations (NGOs) mentioned in this chapter and the rest of the book are listed in Appendix 3, and British zoos housing registered Amur tigers are given in Appendix 4.

The final chapter also considers tiger conservation in the Russian Far East. The Russians have already scored some notable successes over many years - since 1936, when Lev Kaplanov went to work in the Sikhote-Alin Zapovednik (see chapter 7). A wealth of information has been gathered, and is still being gathered, now often with the help of foreign NGOs. It must be emphasised, however, that there is considerable uncertainty about the population estimates, with different survey methods sometimes yielding markedly different results. There is also great uncertainty about prospects for the Amur tiger in the wild - particularly concerning the prey base and poaching pressure. Several decades from now the tiger may be thriving, or it may be extinct - or somewhere in-between; we simply do not know at this stage. Almost certainly, though, serious and realistic conservation measures based on sound knowledge of the tiger's biology will markedly improve its chances. While supporting efforts to save the Amur tiger, we are principally concerned in this book to present information to enable you, the reader, to come to your own conclusions about the situation and, perhaps, to choose one or more of the Amur-tiger organisations to support (see Appendix 3).

If this book had been published several decades ago, it may well have carried a title along the lines of *Tigers Behind the Iron Curtain.* Not so long ago, the Soviet Union had *two* tigers. The other one - the one this book is *not* about - is the Caspian tiger. The success the Russians have so far had in conserving the Amur tiger was perhaps, in part at least, due to their wish not to repeat the sad story of its western cousin.

The Caspian tiger

As chapter 9 makes clear, the Caspian tiger's range is - or was - at the far western end of the species' range, just as the Amur tiger occupies the eastern extremity. Originally it was found in a good part of what became Soviet Central Asia, especially along rivers. Here its habitat was the dense *tugay* vegetation of poplar and willow woodland and tall reed beds. To cope with the vegetation the tiger was said to stand on its hind legs to get a view over the tall grasses. Having to endure fairly cold winters it had a thick winter coat, which was somewhat redder than the Amur tiger's, with long belly fur and beard, and with stripes set closer together.[2]

Tigrovaya Balka was one of the former Soviet Union's nature reserves, though it is now in independent Tajikistan, on the border with Afghanistan. One of the aims in setting up the reserve was to protect the fast disappearing *tugay* vegetation, and therefore also the Caspian tiger's habitat. In the 1930s and '40s the tiger was still common in the area.

'*Balka*' means 'dried up river channel', so the name of the reserve means 'tiger former-river-channel'. The name was acquired following an unusual incident in 1928. In December of that year two border guards on horseback were crossing a snow-covered *balka* when a tiger suddenly leapt onto the horse of the rear guard, who was called Kukuy, and knocked it to the ground. Hearing a commotion, the other guard, Morgun, looked back and saw the tiger standing over the fallen horse and rider. Before Kukuy could reach for his rifle, the tiger slipped away. Leaving the horse, which was seriously wounded in the rump, Kukuy went back but saw that the tiger was following him. At the next *balka* the tiger disappeared into the vegetation. Just then some more guards appeared, who had been fetched by Morgun. Since Kukuy was unharmed they

Caspian tiger
Photographed in Berlin Zoo, c.1900

carried on to the scene of the attack, but it was not long before the tiger jumped onto another horse. As before it seized the horse by the rump, but it also hit its head on the saddle and caught the guard's clothing in its claws. The tiger immediately made off. It all happened so quickly that, once again, no-one had time to fire. The tiger was not seen again. The first horse had to be shot, and the second spent a month or so in a veterinary clinic to recover. From then on, the former river channel was known as Tigrovaya Balka.[3]

Sadly the Caspian tiger represents failure - not *its* failure, but *our* failure to live alongside it in harmony. A few may have been alive as late as the 1970s, but none has been seen since. The Caspian tiger represents a failure in nature conservation, and more generally in the relationship between humans and nature. Reasons for the Caspian tiger's disappearance will be mentioned later in the book. The main subject of this book is, however, the Amur tiger, and that represents much more of a success story - at least, so far. Other chapters will explain why that is so, and why we can have grounds for guarded optimism about its future.

Chapter 2

AMBA [1]

The next day a dense fog enveloped the whole area, making it damp and chilly outside. While the men collected their belongings and loaded up the horses, Dersu and I drank our tea, stuffed rusks into our pockets, and went on ahead. It was my custom to leave camp before the rest: surveying slowed me down, and in two hours or so the detachment would overtake me.[2] I would catch them up during the rest hour, when they had already eaten and were getting ready to set off again. The same would happen in the afternoon, and I would arrive in camp only at supper time.

Dersu had mentioned to me the day before that there were many tigers in this area. He advised me not to get separated from the detachment.

Our route passed along the right bank of the Li-Fudzin.[3] Sometimes the path went off to the side, plunging into the forest so that it was difficult to find our bearings and know where the river was. But then we would suddenly find ourselves by the water once more, and could continue along the steep bank.

No-one who has not been in the Ussuri taiga[4] can imagine what a tangled jungle it can be. You literally see nothing more than a few paces ahead. Just five yards or so away you are likely to startle an animal from its lair, and only the rustle and crack of breaking twigs will indicate which way it has gone. We had been fighting our way through this kind of forest for two days now.

The weather was not kind to us. There was constant drizzle. We waded through puddles on the path. The foliage was soaking, and large drops fell from the trees. An amazing deathly silence filled the forest. Even the woodpeckers had disappeared.

'Devilish weather!' I remarked to my companion. 'If it's not raining, it's fog. What do you think, Dersu? Will the weather clear up, or just get worse?'

The Gol'd[5] looked up at the sky, then around him, and went on without saying anything. A minute later he stopped and said, 'My people think land, mountain, forest - they all people. They now sweating. Hear!' - and he listened intently - 'Hear them breathe. They - all people.' He walked on, and described at length his conception of nature, in which all things were alive, as if they were people.

It was already eleven o'clock in the morning. The detachment should have overtaken us long before this, but nothing could be heard behind us in the taiga.

'We must halt!' I said to Dersu.

Without saying anything he stopped, took his rifle from his shoulder and leaned it against a tree, stuck his bipod[6] into the ground, and began searching for his pipe.

'Tfu! Me[7] lose pipe,' he exclaimed in annoyance.

He wanted to go back and look for his pipe, but I advised him to wait - in the hope that the others, coming along behind, would find it and bring it. We waited about twenty minutes. The old man was clearly dying for a smoke. At last he lost patience, picked up his rifle, and said,

'Me think pipe near. Must go back.'

Anxious about the prolonged absence of the pack horses, and imagining that something must have happened to them, I went back with Dersu. The Gol'd went ahead. As always he shook his head and carried on a running discussion with himself.

'How could me lose pipe? Either me get old, or my head gone queer, or ...?'

He did not finish the sentence, but stopped in the middle of a word, stepped back, stared at the ground, and began examining something at his feet. I went up to him. Dersu looked around, seemed somewhat confused, and whispered,'Look, Captain. Amba. He come along behind. This bad. Spoor very fresh. He - just here.'

It was true. Quite fresh footprints of a large cat could be distinctly seen on the muddy path. When we came past earlier there were no prints. I clearly remembered that - and Dersu would hardly have passed them without noticing. Now, as we headed back the way we had come, here were the footprints, coming towards us. The animal had obviously been right on our heels.

'He very close, he hide here,' said Dersu, pointing to the right. 'He stand here long time, when we stop, look for pipe. We go back, he quickly jump over there. Look, Captain, no water yet in spoor....'

True enough, despite the fact that there were puddles everywhere else, there had been no time for water to trickle into the tiger's footprints. There could be no doubt about it: the terrible beast had been standing here just a moment earlier and then, hearing us coming, had disappeared into the undergrowth and hid itself somewhere.

'He not far. Me know him. Captain, wait ...'

For several minutes we stood stock still in the hope that some sound would reveal the tiger's presence. But the silence was absolute, and somehow especially mysterious and terrifying.

'Captain', Dersu turned to me, 'now we must look very carefully. Your rifle loaded? We must go very quietly. Look every hole, every fallen tree, look very carefully. No hurry. It is Amba. Understand? Amba!'

Saying this, he stared hard at every bush and tree, and so we crept along for about half an hour. Dersu went ahead and never relaxed his gaze.

At long last we heard voices: one of the Cossacks was cursing a horse. Within a few minutes they came up with the horses, two of them covered in mud. The saddles too were filthy. It seems that, when crossing a stream, both horses had stumbled and got stuck in a bog. Hence the delay. As I had suspected, the riflemen had found Dersu's pipe on the path and brought it along.

Before setting off again we had to adjust the horses' loads and clean some of the mud off the animals. I wanted to stop and brew tea, but Dersu recommended that we spend as little time there as possible and get a move on. He said that a hunter's cabin was not far off: that would be a better place to make camp. After thinking about it, I agreed.

The men began to adjust the loads on the exhausted horses while Dersu and I set off along the path once more. We had gone only two hundred paces when we again came across the tiger's track. It had once more been following us and, also as before, had avoided an encounter when it heard us coming. Dersu stopped and, turning in the direction the tiger had gone, shouted with a hint of indignation:

'What you doing behind us? What you want, Amba? We go along path, no hurt you. Why you come along behind? No enough space in taiga for us all?'

He brandished his rifle in the air. I had never seen him so agitated. In Dersu's eyes I could see that he firmly believed that the tiger - Amba - heard and understood everything he was saying. He was certain that the tiger would either take it as a challenge, or leave us in peace and go off somewhere else. After waiting about five minutes, the old man sighed and lit his pipe. Throwing his rifle over his shoulder, he confidently strode off along the path. His face once more became expressionless: he had shamed the tiger, and forced it to go.

For another hour we continued on our way through the forest. Suddenly it began to open out, and before us appeared a large clearing. The path struck across it diagonally.

Our protracted sojourn in the taiga was exhausting us. The eye sought relief and some open space. So you can understand the joy we felt on leaving the closed forest and seeing the open glade.

'This - Kvandagou,' said Dersu. 'Soon we find hut.'

The open space we were now crossing was covered in low bracken.[8] Beyond the forest to the north could be seen, faintly through the mist, the outlines of high forest-covered mountains. On lower ground stood scattered trees - mostly maple, oak and Dahurian birch. On the right side of the clearing was a narrow marsh and salt licks, which, according to Dersu, were visited every night by *izyubr*[9] and roe deer to feast on the water crowfoot and nibble at the salty black soil.[10]

'We must hunt today,' said Dersu, pointing with his bipod at the marsh.

We did indeed find the gable-roofed hut by about three o'clock. It had been built of pine bark by Chinese hunters in such a way that smoke from a fire inside was diverted to both sides, so preventing mosquitoes from entering. A small stream flowed past the hut. Once more, it took a long time to prepare the horses on the far bank, but eventually they got across.

Meanwhile the weather was still, as Dersu had described it, 'sweating'. But the overcast sky had been making a bit of an effort to look brighter; the fog rose, here and there appeared gaps in the clouds and the rain stopped, but the ground was still wet.

I decided we would stay the night here. I had an urge to hunt at the salt licks, especially since for a long time we had eaten no meat: for four days we had been surviving on rusks.

Within a few minutes the camp was a scene of cheerful bustle - well known to anyone who has been a long time in the taiga leading a nomad's life. The horses were unloaded and released. As soon as the saddle was taken off, each horse rolled on the ground, shook itself, and wandered off into the clearing to feed.

In case it rained, the packs were brought together and covered with a tarpaulin. While all this was going on, someone managed to get a fire going and put the kettle on.

In camp Dersu always worked with tremendous energy. He ran from tree to tree, stripped off birch bark, cut poles and stakes, put up the tent, dried his own and everyone else's clothes, and tried to arrange the fire in such a way that it would be possible to sit in the hut without smoke getting in our eyes. I was always amazed how this old man managed to do several things at once. Even when, with boots off, we had been relaxing for some time, Dersu was still dashing around keeping very busy.

An hour later an observer watching us would have beheld a curious scene. The horses, their backs wet with rain, are grazing in the clearing near the stream. The smoke from the fire is not rising, but stays close to the ground, apparently motionless. Escaping from the mosquitoes and midges, the men are out of sight inside the hut. Just one person is still running around among the trees: this is Dersu, feverishly collecting firewood for the night.

On an overcast August day, it gets dark early. This day was no exception. Cloud capped the tops of the hills and patches of mist drifted through the bushes, like ghosts.

As soon as we had finished supper, Dersu and I went off hunting. We took the path which had brought us to the hut, and then cut across to the salt licks near the forest. All around the marsh was a multitude of *izyubr* and roe deer spoor. The dark soil of the salt licks was almost completely devoid of vegetation. Small trees around them looked stunted and sickly. In places the ground was well trampled. *Izyubr* clearly came here frequently, both singly and in groups.

Choosing a convenient spot we sat down and waited. Leaning against a stump I began to watch. It rapidly grew dark among the bushes and under the trees. For a long time Dersu could not settle down: he broke off branches, in order to get a clear view for firing, and for some reason bent down a young birch tree growing behind him.

All around in the forest and in the open reigned a deathly silence. It was broken only by the monotonous buzzing of the mosquitoes. This kind of silence is somehow particularly depressing. You involuntarily get drawn into it, dominated by it, and you lack the strength to disturb it, even just by saying something or making a careless movement.

The light continued to fail. Bushes and trees lost their shapes. They seemed to come alive, and gave the impression of moving about. Sometimes I fancied they were deer, and my imagination filled in the missing details. I would raise my rifle, ready to fire, but each time I would see Dersu's calm features and that would bring me back down to earth. The illusion immediately disappeared, and the dark silhouette of a deer would turn back into a tree or bush. Dersu sat there like a marble statue. He attentively watched the bushes around the salt licks, and calmly waited for his quarry. On one occasion he suddenly became alert, quietly raised his rifle and began to take aim. My heart was pounding. I stared in the same direction, but saw nothing.

Soon noticing that Dersu was calm once more, I also relaxed. It became pitch black - so dark that, just a few paces away, it was quite impossible to make out either the dark soil on the salt licks or the shapes of the bushes. The mosquitoes were unremitting in their attacks on my neck and arms. I covered my face with a net. Dersu sat there without a net and appeared to be oblivious of the insects.

Suddenly I heard a rustle. This time I was not imagining it. The sound came from bushes beyond the salt licks, directly opposite us. I glanced at Dersu. He was craning forward and seemed to be straining with all the power of his eyes to penetrate the darkness and discover the cause of the noise. Sometimes the rustle grew louder, becoming exceptionally clear, but then would die away and cease altogether. There could be no doubt that something was creeping towards us cautiously through the vegetation. It had to be an *izyubr* coming to nibble and lick the salty ground. In my mind's eye I already saw the elegant deer with magnificent branched antlers. I pushed up my veil to listen and stare, ignoring the mosquitoes. I reckoned that the deer must be no more than seventy or eighty yards away.

Suddenly a fearful growl, like distant thunder, came through the air: 'R-r-r-r-r-r!'

Dersu seized my arm. 'Amba, Captain!' he gasped in fright.

Terror overcame me. It is not easy to describe what I felt. It was as though a lassitude, or heavy load, began to weigh my legs down. My knees ached, just as if lead had been poured into them. The feeling is unforgettable by anyone who has suddenly been scared stiff by something. But at the same time another sensation filled my heart - a mixture of curiosity, veneration for the terrible regal beast, and excitement of the hunt.

'Bad! We should not be here. Amba - angry! This place - his,' said Dersu - and I still do not know whether he was addressing me. I had the impression that he was scared out of his wits.

' R-r-r-r-r-r!' - again resounded through the still of the night.

Suddenly Dersu rose to his feet. I expected him to fire, but was amazed to see that his hands were empty. Instead he was talking to the tiger:

'All right, all right, Amba! No be angry, no need! This - your place.We not know. We go some other place. Taiga - big enough. No be angry!'

The Gol'd was standing, his arm stretched out towards the tiger. Suddenly he dropped to his knees, twice bowed to the ground, and in an undertone began speaking in his own language. For some reason I felt pity for the old man. Then Dersu slowly got up, went over to the stump, and picked up his rifle.

'Captain, we go!' he announced and without waiting for an answer set off through the undergrowth to the path.

Without thinking twice I followed him. Dersu's calm expression, the confidence with which he strode ahead without taking any precautions or looking around, calmed my nerves. I felt confident that the tiger was not coming after us and would not attack.

After about two hundred yards I stopped and tried to persuade the old man to stay a bit longer.

'No,' said Dersu, 'Me can't. Me tell you before, me never shoot Amba in company! Listen, if you shoot Amba, me no your friend...'

He fell silent and strode off along the path. I wanted to remain behind alone, but the idea terrified me. I ran to catch him up.

The moon was rising, so there was a little more light to see by. Far off, at the other end of the clearing, flickered the light from our camp fire: one moment it would flare up, but only to die down again.

For some time we walked on in silence. We each had our own thoughts and recollections. I was sorry not to have seen the tiger. I said so to Dersu.

'No, no!' he replied. 'See him - bad. We say: if man no see Amba, him happy. Him always live well.'

Dersu sighed deeply, fell silent for a moment, and then went on: 'Me see Amba often. One time me fire at him. Now me very afraid. Some time it will be bad for me!'

The old man's words expressed such profound anxiety that I once more felt pity for him. I began consoling him and changed the subject of conversation.

In an hour we arrived back in camp. The horses, frightened at our approach, shied away and snorted. The men were bustling around the fire. Two of the cossacks came to meet us.

'For hours the horses have been fidgety,' said one of them. 'They have not been grazing, all the time looking around. Is some animal nearby?'

I ordered them to tether the horses, light a fire, and set an armed watch.

All evening Dersu was silent. The encounter with the tiger had made a deep impression on him. Straight after supper he turned in, but for a long time he could not sleep. He was tossing from side to side, and seemed to be muttering to himself.

I told the men what had happened. The cossacks listened with rapt attention, and then began recalling their own lives on the River Ussuri, their hunting adventures, who had seen what, and who was present on each occasion.

We talked until late. Eventually sleep crept up on us: one man yawned, another made his bed and lay down. In a few minutes everyone was asleep.

Silence reigned once more all around. The only sounds in the hut were the measured breathing of the sleeping men and the crackle of branches on the fire. Outside could be heard the snorting of the horses, the hoot of an eagle owl in the trees, and from very far away the cry of a pygmy owl.[11]

Biographical Sketch of Vladimir Arsenev[12]

Vladimir Klavdievich Arsenev was born in 1872 in St Petersburg. His father was the illegitimate son of a member of the urban lower middle class, but worked his way up to become head of the railway system in Moscow Region. There were eight children in the family, and several of the sons were probably inspired to travel later in life by the books of exploration in the house, as well as geographical games their father would play with them. Vladimir's maternal uncle also inspired the boy to love nature.

Vladimir entered cadet college with a view to joining the army - this being one way to see the world. While at school he was taught by the well known explorer M.E. Grum-Grzhimaylo, who painted an enticing picture of East Siberia and the Russian Far East. On entering the army the young Arsenev first went to Poland, but in 1900 had the chance to serve in the Far East; he was posted to Vladivostok.

Russia's defeat in the 1904-05 war with Japan convinced the Russian Far East authorities that they needed to know more about their own poorly explored mountain country: it might, after all, provide an invasion route in the case of a future Japanese attack. Arsenev's suitability for the job was recognised, and so he found himself exploring the Sikhote-Alin Mountains north-east of Vladivostok. In particular he was required to construct maps, but also to describe the natural resources, native peoples, settlers and settlements.

So began the series of expeditions that became the foundation of Arsenev's reputation. The most important were in the Sikhote-Alin Mountains, but he also travelled extensively in other parts of the Russian Far East.[13] Each expedition would consist of a detachment of soldiers, horses to carry the equipment and provisions, sometimes a botanist or other expert, and a native or Chinese guide. One guide, belonging to the Nanay people, particularly impressed Arsenev, and they developed a close relationship in the nineteen months between their first meeting in 1906 and the guide's death at the hands of robbers. This was the famous Dersu Uzala, who featured prominently in Arsenev's early books.[14] While the books read like diaries, they are in fact partly fictional - at least so far as Dersu is concerned. In the books the two men first meet in 1902, Dersu's wife died earlier in an epidemic, and he had no other family; in reality they did not meet until 1906, and Dersu did in fact have close relatives, including a brother. It appears that, to give his books more of a literary character, Arsenev somewhat idealised Dersu, so that he became almost the Russian equivalent of the wise Red Indian of North America. Dersu Uzala came to world attention with the award-winning film of that name by the Japanese director Kurosawa.

Vladimir Arsenev (1872 - 1930)

When not on expeditions Arsenev held various civilian jobs, led the life of an academic, and wrote technical reports and the popular books he is famous for. From 1910 to 1918, and again from 1924 to 1926, he was director of the Regional Studies Museum in Khabarovsk. Between the February and October revolutions in 1917 he was Far East Nationalities Officer for the Provisional Government. Later he was working for a fisheries trust. From 1926 to 1930 he lectured at the Far East State University in Vladivostok.

Arsenev and his family could not expect to have an easy life during such turbulent and revolutionary times. He appears to have welcomed the February Revolution, which removed the monarchy. On the other hand he did not approve of the October Revolution, which brought the communists to power in St Petersburg - though not until 1922 in the Russian Far East. From 1923, as a former tsarist army officer and then official of the Provisional Government, Arsenev was required to report to the communist authorities once a fortnight, but otherwise he was mostly left in peace. In 1918 his first major tragedy struck: his parents and other members of his family back in Ukraine were murdered by bandits. In the same year Arsenev divorced his first wife and married Margarita Nikolayevna Soloveva - whose father, incidentally, subsidised publication of *Dersu Uzala* four years later.

By 1928 he was apparently well in favour with the political authorities, since in that year a letter from Maxim Gorki, the well known author and playwright, was published praising Arsenev's literary work. However, his position was always vulnerable; certain people were trying to undermine it, and in 1926 he was summoned to answer accusations. Although his arrest was ordered in 1930, he died in 4th September before it could be carried out.

In 1934 however his widow Margarita Nikolayevna was arrested, but was released after a few months for lack of evidence. Three years later she was arrested again, accused of being a Japanese spy, and this time was shot. In 1939 Arsenev's daughter, not knowing that her mother had been executed, was herself briefly arrested on a morality charge. Thefollowing year she was arrested for anti-Soviet propaganda and was sentenced to ten years in a labour camp. In 1970 she died at the early age of forty-nine.

Meanwhile Arsenev was rehabilitated in the 1940s, and much later his remains were transferred to a prestigious cemetery. He was given the status of iconic poet-explorer, and so he remains today.

View across the river Bikin, with hills rising above the distant mist. Occasional emergent Korean pines can be seen in the forest on the far bank.

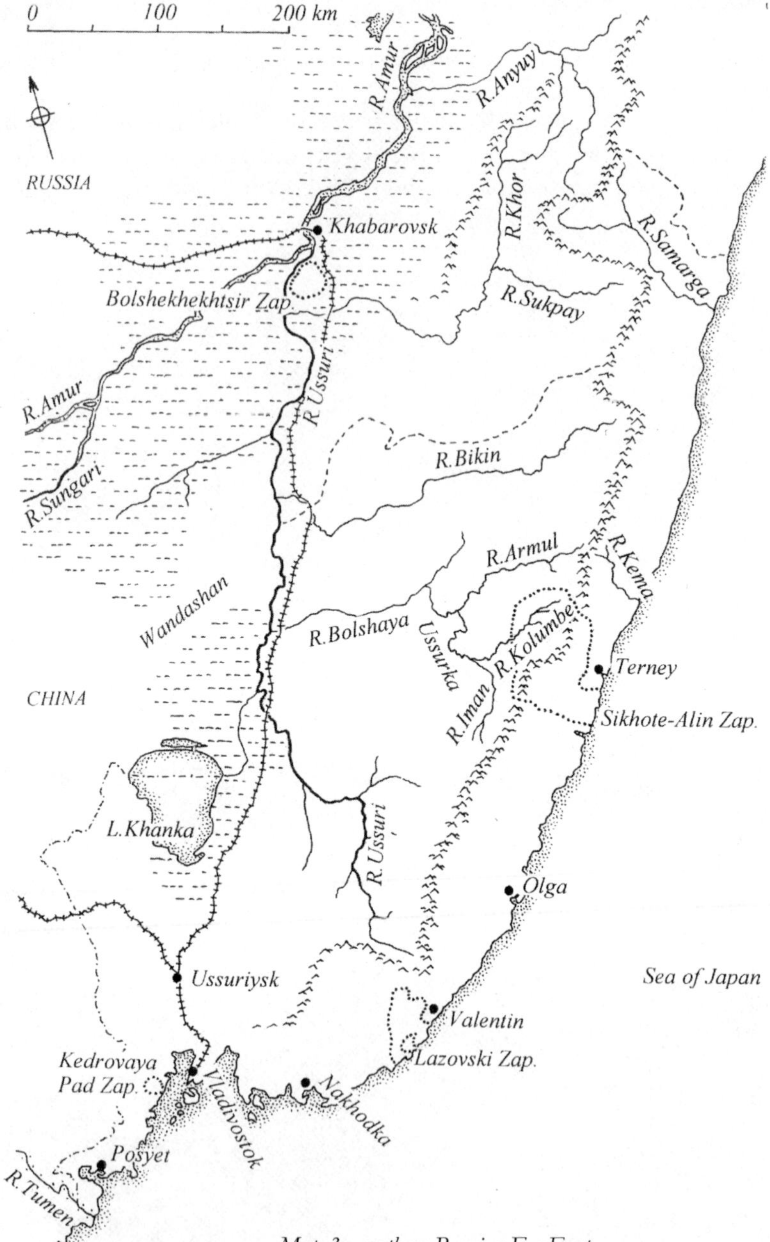

Map 3: southern Russian Far East

Mountain symbols - the main Sikhote-Alin watershed and a smaller one west of the Khor. Horizontal dots - extensive lowlands around L. Khanka and the main rivers. Dashed line - boundary between Khabarovsk Kray to the north and Primorski Kray (Primore) to the south; between the headwaters of the Bikin and Samarga it passes along the main watershed. Four zapovedniks are bordered by lines of dots.

Chapter 3
RUSSIAN TIGER COUNTRY

Moscow is at nearly 56°N latitude - about the same as Edinburgh, where this chapter is being written. Imagine flying due east from Moscow until you come to the sea. You would cross 100° of longitude, each about 63 km, so that the total distance would be not far off 6300 km (see Map 1, p.ix).

Leaving Moscow you would pass over fairly flat country until you came to the Ural Mountains. On the way you would cross the great River Volga, which flows south to the landlocked Caspian Sea. The Urals are not particularly high, and to their east the flat land continues. Flowing north to the Arctic Ocean are other mighty rivers - the Irtysh, Ob and Enisey[1]. This stretch from the Urals to the Enisey is the West Siberian Plain. Further to the east the country is more mountainous, and in the hills just west of Lake Baykal you would spot a stream which further north becomes a giant among Arctic rivers - the Lena. Again more mountains, until eventually they drop away and the Sea of Okhotsk appears below you. This sea is almost landlocked, with the Kamchatka peninsula to the east, Sakhalin and the mainland to the west and north, and the volcanic Kuril Islands and Hokkaido to the south.

Non-Russians commonly think of Siberia as all Russia east of the Urals. For Russians, however, Siberia is not quite so extensive. East of the mountains beyond Lake Baykal, the lowlands associated with the Pacific coast and its rivers have historically been regarded as a more or less self-contained area - the Russian Far East (or, formerly, the Soviet Far East). Its formal boundaries have shifted at various times through history, but they are now more or less as shown on Map 1. Due east of Moscow it forms quite a narrow coastal strip, 250-400 km wide, but to both north and south it broadens out considerably. Since tigers have been confined, at least in historical times, to the far south, mostly south of Khabarovsk, the following description will concentrate on that area.

Geography and climate

This part of the world is a geographer's paradise. The west and north of the Pacific are bounded by graceful sweeps of islands and peninsulas. One is formed of Honshu, Hokkaido and Sakhalin. From Hokkaido sprouts the Kuril Islands chain, continuing part way up the Kamchatka peninsula as several lines of volcanoes. At the northern end is the tallest of them - Klyuchevskaya Sopka, 4750 m high. With another change in direction, a chain of islands - comprising Russia's Komandor Islands and the United States' Aleutian Islands - crosses the Pacific to Alaska. On the seaward side of these 'island arcs' are trenches in the sea floor, nearly 10,000 m deep in places. They represent where the floor of the Pacific Ocean is sliding under the edge of the continental crust, and creating the volcanic activity found in many parts of this Pacific 'ring of fire' [2] (see Map 2, p.x).

Only in 1849 was it confirmed that Sakhalin is in fact an island. It is almost joined to the mainland near the mouth of the River Amur. The Amur drains southern Siberia east of Lake Baykal and part of northern Mongolia before zigzagging its way east. The rivers Argun and Amur form the border between Russia to the north and China to the south, at least as far as Khabarovsk. At this point a major tributary - the River Ussuri - joins the Amur from the right, and the Russian-Chinese border continues south-west along it nearly to Lake Khanka. Here it crosses the northern part of the lake before running about 400 km through the hills to join the coast where Russia meets North Korea.

The sea would have to fall only 50 m for Sakhalin to be joined to the mainland. On the other hand, if it rose less than 100 m it would create an island out of all the land east of the Ussuri and Amur rivers below Lake Khanka. This is the mountain area known as the Sikhote-Alin.[3] Its highest point is to the east of Khabarovsk - a peak just over 2000 m high. Tigers formerly ranged through most of these mountains, and also much more widely. For instance they were abundant in the Manchurian hills west of Lake Khanka, and also in the extensive reedbeds around the lake - a habitat not unlike that of the Caspian tiger[4] in Tajikistan. However, tigers are now largely confined to the Sikhote-Alin south of Khabarovsk. The tiger's range is described in chapter 9.

Map 3 shows the area. Horizontal lines of dots mark the flat and originally swampy land near Lake Khanka and the large lowland rivers. Between the Ussuri and the sea, the Sikhote-Alin Mountains are drained to west and east of the main watershed, marked as a line of peak symbols. For most of the chain's length, the watershed is much closer to the sea than to the Ussuri. This means that the rivers flowing to the Ussuri have further to go than the eastward rivers - a feature enhanced by their frequently meandering courses. So the rivers of the western macroslope (meaning the area draining to the west) tend to have a gentle gradient, and the climb to the main ridge from that side may involve few really strenuous ascents except near the ridge itself. On the eastern macroslope, by contrast, the rivers have less distance to reach the sea, and often go in an almost straight line. As a result of their steep gradients these rivers are often tumultuous torrents. This physical shape of the landscape alone gives the eastern and western macroslopes of the Sikhote-Alin quite different characters.

The difference in landscape is accentuated by the climate. The whole area falls in the Temperate Monsoon Zone, and that means Arctic air in winter and temperate air in summer. In winter there is often an area of high pressure situated over Lake Baykal or some other part of Siberia, and the result is that northerly winds bring dry and very cold air. In summer, by contrast, pressure over the continent is low and so winds generally blow off the sea from the south, bringing cool moist air. The cooling effect is also enhanced by a cold current flowing south along the coast. Warm moist air being blown onshore across the cold current produces frequent summer fogs.

Winters therefore are generally clear, sunny, cold and windy, with rather little snow; but if the wind comes from the south, it usually brings blizzards. Sea ice appears most years along the coast north of Vladivostok. After a long cold spring the summer pattern of weather is normally established by early May, and lasts until September: the generally light winds bring abundant rain, and frequent fogs in spring and early summer. Summer days may be calm, but they are often overcast, and the humidity is high. As a result of the small amount of snow, and most of the rain coming in summer, the rivers tend to flood in summer - rather than in spring, which is usual in European Russia.[6] Typhoons occur in summer and autumn, and these can cause catastrophic floods, especially where hill slopes have been deforested. On the eastern macroslope and along the coast, autumn is warm and prolonged.

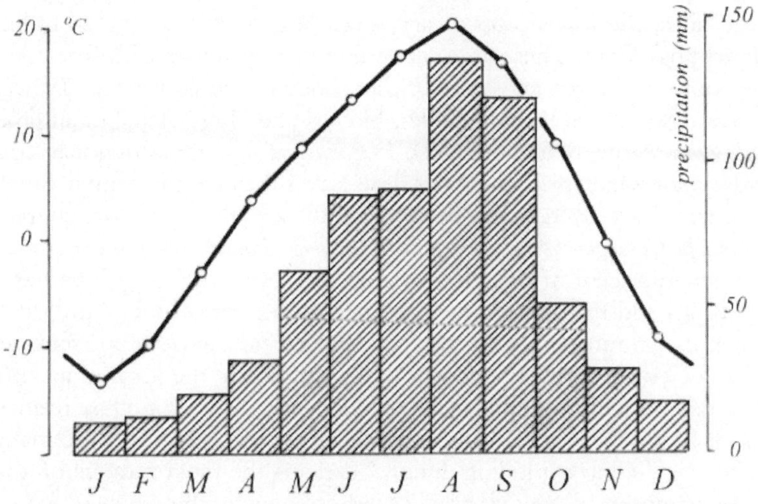

The climate at Vladivostok[7]

The line graph and left scale show temperature month by month. On average it is well below zero from December to March, but then rises rapidly to a peak in August. The columns show precipitation (rain and snow), and produce a remarkably similar shape - meaning a wet summer but fairly dry spring and autumn. There is more rain and snow in the hills.

There is considerable climatic variation in different parts of the Sikhote-Alin. The further a place is from the coast, the earlier summer begins and ends, and winters are colder.[7] Moreover precipitation - in other words, rain and snow - increases by about 20% for every 100 m increase in altitude. So while, even in summer, the coast may have rather few days with rain, in the mountains it can be wet nearly every day. Because the moist air off the sea is forced to rise so rapidly up the steep eastern slope, a great deal of its moisture is shed on the eastern macroslope; by contrast the western macroslope tends to have more gentle rain. The altitude effect also explains why in winter there is much deeper snow in the mountains than in the lowlands or on the coast.

Vegetation

In Arsenev's time, most of the Sikhote-Alin Mountains were still covered in virgin forest - except for water-logged valley bottoms and the stony wastes above the tree line. He noted that 'In the Kema valley grows a splendid primaeval forest, never touched by human hand. It was as if Nature had deliberately selected these places to show the creative power of the soil'.[8] These forests were the main habitat of the Amur tiger.

As usual in mountains, the vegetation varies considerably with altitude. In the Sikhote-Alin range, however, this altitudinal zonation is of a special kind. Apart from the willows, poplars and other trees found along the river sides, the lower slopes carry deciduous forest of the so-called Manchurian flora. These are species that have affinities with forests to the south, in Korea and China, where the climate is warmer. Typical species are various hazels, maples, birches and elms, Manchurian lime *Tilia mandshurica*, Manchurian ash *Fraxinus mandshurica*, Mongolian oak *Quercus mongolica* which retains its dead leaves through the winter, Amur cork tree *Phellodendron amurense* with its thick corky bark, Manchurian walnut *Juglans mandshurica*, the exotic *Kalopanax ricinifolia* with its enormous leaves reminiscent of the tropics, and *Aralia elata* with its vicious spiny stems. To add to the exotic appearance of the forest, there are lianes which can climb high into the trees and eventually develop stems as thick as a man's arm. Prominent among these are the local grapevine *Vitis amurensis*, the kolomikta vine *Actinidia kolomikta* and its close relative *A.arguta*, and the lemon-scented *Schisandra chinensis*. There are also many shrubs, including exploding dittany.[9] Perhaps the most famous of the herbs is ginseng *Panax ginseng* - eagerly sought by the Chinese for its medicinal roots.[10]

Climbing higher up the slopes, you come into quite a different kind of forest. It is dominated by conifers, in particular Yedo spruce *Picea ajanensis*,[11] Khingan or East Siberian fir *Abies nephrolepis* and the larch *Larix gmelinii*. Apart from the deciduous larch - mostly found further

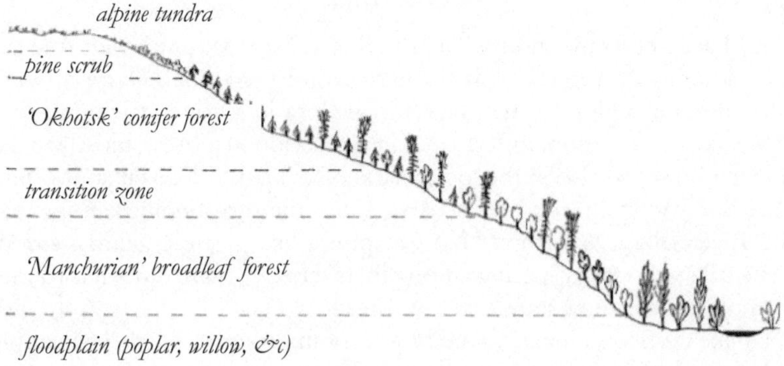

Simplified vegetation profile

The deciduous woodland on the lower slopes gradually gives way to conifer forest higher up. In both zones, as well as in the transition zone between them, mature Korean pine trees will often sore above the rest of the canopy. Some of the poplars in the valley bottoms can also attain great heights.

north - these are evergreens, which can form a dense canopy allowing little light through. The Russians call such forest 'dark-coniferous taiga'. In contrast with the Manchurian flora comprising the low-altitude forest, the dark-coniferous and larch forests are made up of species from the so-called Okhotsk flora. Most of its species have ranges extending far to the north, and indeed are typical members of the Siberian taiga.

There is a striking contrast between the dark green conifers on the upper slopes and the deciduous woodland below - especially when the autumn colours are at their best, and later when the trees lower down have lost their leaves. Instead of the alpine meadows found above the tree line in most mountain ranges, in the Sikhote-Alin there is a belt of pine scrub *Pinus pumila*, which can be very difficult to walk through or over. Higher still is alpine tundra formed of small shrubs and mosses, much as on the coast of the Arctic Ocean.

This juxtaposition of forests of the southern Manchurian and northern Okhotsk floras creates an extraordinary situation in the Sikhote-Alin. Animals and plants characteristic of the Siberian taiga penetrate almost as far south as Vladivostok, though progressively higher up the mountains the closer they are to the southern end of the range. Elements of the southern forests penetrate as far north as about Komsomolsk-on-Amur, but are progressively confined to the lower altitudes, and with fewer and fewer species towards the north. So, throughout the southern Sikhote-Alin, 'southern' animals such as wild pig, *izyubr* and sika deer, Asiatic black bear, raccoon dog and the Amur tiger can be found in close proximity or even intermingling with 'northern' animals like the elk, wolverine, wolf, brown bear, lynx and sable.

We have yet to mention what is perhaps the most important tree of the Sikhote-Alin forests. This is the Korean or red pine *Pinus koraiensis*. Although belonging to the Manchurian flora it is not restricted to the lower slopes: it is found intermixed in the deciduous forest, mostly in its upper zones, and also in the lower parts of the coniferous taiga. It often occurs as widely spaced trees soaring above the surrounding canopy: on the River Bikin, 'enormous Korean pines, like gigantic guards, stood quite still without whispering among themselves'.[12] Sometimes it is found as more or less pure stands of tall forest.

Unlike the Scots pine *Pinus sylvestris* so common across northern Europe and western Russia, the Korean pine does not have its needles in pairs, but instead in bunches of five. If the needles were much shorter, the tree might bear a passing resemblance to a cedar, and so the Russians call it '*kedr*' or 'cedar'.[13] A far more important difference, in ecological terms, is that its cones contain up to 150 tasty nutritious nuts - instead of the winged and wind-dispersed seeds of the Scots pine. As will emerge later in this book, the fates of many of the Sikhote-Alin animals, including the tiger, depend on the pine nut. This would be true if the pine produced a good crop every year, and various mammals and birds came to feed on it every autumn. However, there may not be a good crop every year, so

Korean pine

Mature trees - up to 40 m in height - are rather untidy since the trunk tends to divide high up to give many leading shoots, and sometimes a broad flat top. The whole scene is forest-covered except for the tops of the mountains in the far distance. (Based on a photograph by the author.)

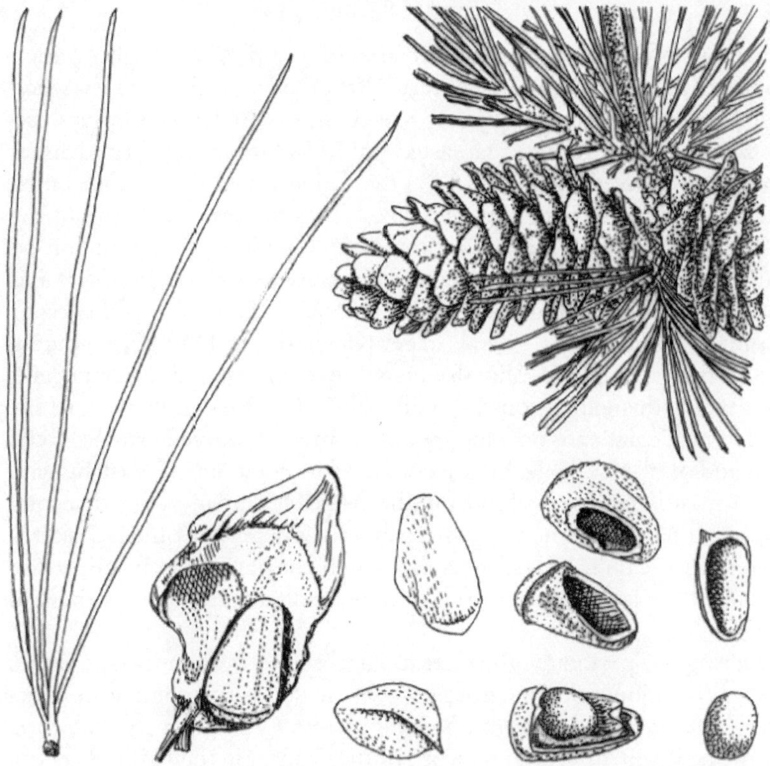

Korean pine[17b]

Each short shoot (left) has five needles. The cone is c.15 cm long, and its thick scales hold the nuts in depressions on the inner surface. When the shell of the hard nut is cracked or nibbled away, the edible kernel falls out. Besides the nuts, Korean pine produces good-quality timber, and so has been a prime target for logging companies. (Drawn life size, except for the cones which are about one third life size.)

that the animals have to adapt to the pine crop each autumn, according to how heavy it is. Woe betide them if the pine crop fails.

Human history[14]

From time immemorial there have been native peoples living in the area - the Nanay, Udege and others. Chapter 5 returns to them, and describes some of their lore about tigers.

They now play a minor role in the Russian Far East. The reason is not hard to find: 'Immigrants from Europe and Asia created a mosaic of Chinese, Korean, Japanese, Russian, Ukrainian, Cossack, Polish, German, Finnish, and Estonian communities'.[15] The history of the Russian Far East is really one of colonisation. As early as the 7th century AD Chinese imperial power was extended into the area, and Koreans also settled southern Primore in the 8th century.[16] From the mid 13th century, when the Mongols were conquering the Russians in Europe, they also dominated Primore, and that lasted until 1368 when the Yuan dynasty collapsed. Between 1413 and 1433 the Chinese sent expeditions down the Sungari and Amur rivers, almost as far as the sea.

The mid 17th century was a remarkable period of exploration by Russians, with accompanying expansion eastwards through Siberia and the Far East. The Amur valley was explored during the expeditions of Vasili Poyarkov in 1643-46 and Erofey Khabarov[17] in 1649-53, but during this period the Sikhote-Alin Mountains were not entered. An important motive for this expansion was 'soft gold' - the Russian name for furs, and in particular those of squirrel, otter, marten, beaver, ermine, arctic fox, and of course sable. Furs accounted for about one third of Russia's state treasury receipts in 1660, but the insatiable demand was exhausting supplies on the Urals and in western Siberia. There were other subsidiary motives for expansion, such as the search for defensible frontiers and the need for land, partly to settle religious dissenters, such as the Old Believers.[18]

During this period the Russians usually built forts at important places, such as river junctions and portages - where goods and sometimes boats were carried overland from one river system to another. Russian forts along the River Amur were attacked by the Chinese in the 1650s. Realising that establishing themselves in the area was not going to be easy, the Russians signed the Treaty of Nerchinsk with the Chinese in 1689, giving up claims to the Amur in exchange for establishing trade relations.

It was only in the mid 19th century that Russians once more took an interest in the Amur area. They began arriving in small numbers, and this time Chinese opposition was weak on account of political problems in China. Nikolay Muravëv, appointed governor-general of East Siberia in 1847, set about establishing Russian power in the Amur valley and further south. Taking advantage of Chinese weakness, and also the Crimean War, he sent out several military expeditions during the 1850s. By 1860 Vladivostok had been founded, and Muravëv had claimed for Russia all the country between the River Ussuri and the sea, including the coastal strip as far as Korea, so cutting that part of China off from the sea. With British and French forces in Beijing in 1860, China was in such a weak position that it had to agree to Russia's demands, and the borders which Russia had imposed were confirmed in the Treaty of Beijing of 14th November 1860. For his services Muravëv was granted the privilege of being called Muravëv-Amurski.[19]

The region was not actively developed during the next two decades. In about 1880 however it entered its next phase, with a Priamur governor-generalship being established independent of East Siberia, as a response to increasing Chinese and Japanese self-confidence and escalating Anglo-Russian rivalry. The Russian Far East was linked with the rest of Russia by building a railway line, with Chinese agreement, across Manchuria through Harbin. Building commenced in 1897. Military units were responsible for security along the line, and became known as 'Matilda's Guards'. They were joined in 1901 by a young officer, Nikolay Baykov - the author of the story in chapter 10. He loved the Manchurian mountains and forests, and was a skilful naturalist, explorer and ethnographer. He came to an understanding with Arsenev that they would explore respectively Manchuria and the Sikhote-Alin, without poaching each other's 'territory'.[20] The Chinese Eastern Railway to Vladivostok was completed in 1903. It was only in 1916, when a bridge over the Amur at Khabarovsk was opened, that the Trans-Siberian Railway become operational over its full length.

Between 1859 and 1917 over half a million people moved to what was being called 'New America'. By 1911 the native peoples in the Russian Far East comprised only about 5% of the population. European peasants were attracted by land grants of approximately 100 hectares per family. Since most of them settling in Primore between 1882 and 1907 came from Ukraine, the area became known as the 'Ukrainian Far East'. At the same time there were many Chinese - farmers or traders, or roaming the Sikhote-Alin in pursuit of furs, sika and *izyubr* antlers, or ginseng.

The Russian Far East was affected by the Russo-Japanese War of 1904-05, the First World War (1914-17) and the Civil War (1918-22). The history of the Civil War in the area was particularly complex, but one personality stood out from those times - at least when Soviet propaganda had got to work. The name of Sergey Lazo came to prominence when, in 1918, he twice successfully repulsed the anti-Bolshevik forces of Grigori Semënov. In June 1919 Lazo and his partisans attacked 200 American troops near Vladivostok. Then in April 1920, while fighting a Japanese garrison at the mouth of the Amur, he was captured. After being questioned he was taken away by train in a mail bag, then knocked unconscious and thrown into the locomotive furnace. This is how martyrs are created, and Lazo's name will crop up again later in this chapter.

There was another massive influx of settlers between 1931 and 1939 - over a million of them. By 1940 the population of the whole of the Russian Far East had grown to 3.16 millions - still incidentally a very low population density by comparison with parts of western Europe or North America. Even so, it meant that pressure on agricultural land, the forests and wildlife was ever increasing. The influx of people in the 1930s was part of the Soviet government's investment in the region, in anticipation of attack by Japan. There was in fact limited fighting in July 1939 at

Khalkhin Gol on the border between Mongolia and Manchuria - at that time occupied by Japan as the puppet state of Manchukuo - and the Russian Far East went onto a war footing. However, to everyone's surprise, the Hitler-Stalin Pact was signed later that year and a cease-fire was arranged on the Mongolia-Manchukuo border. A Soviet-Japanese Neutrality Pact was signed in April 1941.

Despite the Hitler-Stalin Pact, Germany invaded the Soviet Union in June 1941, and this revived the Japanese threat in the Russian Far East. In the spring of 1942 rumours of an invasion were circulating, but nothing came of them. Then in October 1944, Moscow issued a postage stamp commemorating Lazo - murdered 24 years earlier with Japanese complicity - as a hint that war with Japan was still possible. In fact fighting did not break out before Germany announced its surrender on 9th May 1945. Soviet troops then poured into the Russian Far East with a view to invading Manchukuo, Korea, southern Sakhalin and the Kuril Islands. Japan withdrew from the war on 15th August and in the final settlement the Soviet Union reoccupied the whole of Sakhalin and all the Kuril Islands.

The events of the later 1930s and early '40s, and the corresponding atmosphere and conditions of life in the Russian Far East, have been mentioned in more detail than might be expected in a book about tigers because they set the scene for the tragic events described in chapter 7.

Tiger country

Forested hills - often flat-topped - in the Sikhote-Alin Mountains. Fog commonly fills the valleys. The trees in the foreground are Mongolian oak (left) and Aralia *(right).*

Nothing whatever to do with tigers, however, is the fact that in February 1947 a very large meteorite passed near Khabarovsk and crashed into the foothills of the Sikhote-Alin.[21]

Zapovedniks

In the field of nature conservation, Russia - and from December 1922 to December 1991 the Soviet Union - were and are renowned for their system of zapovedniks.

A zapovednik is a nature reserve set up for scientific purposes, mainly to study ecology. Now numbering at least 100 in Russia, the reserves are intended to represent all major biomes within the country. The term 'zapovednik' derives from a biblical word meaning 'prohibition', and describes accurately enough the intention that nature is to be left undisturbed: only scientists should be allowed in to make observations, and guards to prevent poaching, logging or other damage.[22]

The southern Russian Far East is rather well endowed with zapovedniks,[23] but only five feature in this book. Four are shown on Map 3.

Earliest to be set up was Kedrovaya Pad, close to Vladivostok; the name means 'cedar (ie Korean pine) valley'. Some kind of reserve seems to have been there as early as 1916, but official recognition dates from around 1925. The hills, reaching about 800 m, are covered with oak, lime, maple, and mixed broadleaf forests, with all four main species of liane; there is also some dark-coniferous taiga. The fauna includes both leopard and tiger, as well as the Far East forest cat.

In 1935, mainly at the instigation of K.G. Abramov, the Sikhote-Alin Zapovednik was founded. The original area was enormous - about 1.7 million hectares - and it stretched from Terney as far north as the River Bikin. The whole zapovednik system underwent a catastrophic reorganisation in 1951,[24] during which this reserve was cut back to 99,000 hectares. Subsequently it expanded once more to its present size of about 408,000 hectares, although numerous voices argue that it really needs to be much larger if the Amur tiger is to be saved for posterity.

The present reserve includes both east and west macroslopes of the Sikhote-Alin range, with some peaks up to about 1600 m above sea level. There is a complex network of rivers and streams. Almost the whole area is covered in forest, clearly divided into altitudinal zones. A coastal strip comprises oak with scrub and herbage. Then comes the mixed pine-broadleaf forest, and above that spruce and fir become increasingly dominant. Where conditions are too severe for conifers, the stone birch *Betula ermanii* takes over, and beyond the tree line is *Pinus pumila* scrub and finally alpine tundra. In the valleys are found poplars, willows, *Chosenia* (related to willows and poplars), ash, elm and alder. Over a thousand species of flowering plant have been recorded. In conservation terms the most important mammals are probably the Amur tiger and a curious short-necked antelope, the goral.[25]

In 1937 an area on the coast due east of Vladivostok was set up as a branch of the Sikhote-Alin Zapovednik. Three years later it became an independent zapovednik, with the name Sudzukhinski. Other reorganisations followed, including temporary dissolution, and it is now known as Lazovski Zapovednik, named after Sergey Lazo - who, as far as is known, had no interest in nature conservation. Because Lev Kaplanov worked there as director, his name was added, so that it is now the Kaplanov Lazovski Zapovednik - probably the only Russian reserve named after two people.

Lazovski Zapovednik is about 250 km south-west of Sikhote-Alin Zapovednik, and a third its size. Even so it contains peaks up to 1400 m above sea level, with rivers descending the mountains rapidly and steeply to meet the sea along the picturesque coastline. Most of the area is forested, oak predominantly, but also with patches of birch *Betula mandshurica*, aspen and lime. There is some mixed pine-broadleaf forest. The trees along the valley bottoms are much as in Sikhote-Alin Zapovednik. The flora also comprises about a thousand species of flowering plants, but is noticeably more 'Manchurian' than that of the reserve further north. Mammals include tiger, Far East forest cat, Asiatic black bear, goral and sika deer, and formerly leopard.

Goral

This goat-like animal has small horns and a short thick neck. The tail is quite large, with a black central stripe, which often extends along the spine. The body is grey, or reddish brown.

Bolshekhekhtsirski Zapovednik is a medium-sized reserve, which was set up in 1963 in wooded hills on the south side of Khabarovsk. As recently as 1990 parts of Lake Khanka and its shore were set aside as Khankayski Zapovednik.

Chapter 4

ATTACK! [1]

In the morning I paid off our two guides. It led to a comical incident. I gave each ten roubles - to one a ten-rouble note and to the other two five-rouble notes. The first one was annoyed. Thinking he was dissatisfied with the pay, I pointed out that the other guide was content with his ten roubles.

Evidently I had got the wrong end of the stick. The Udege was annoyed that I had given him just one piece of paper, and his colleague two. I had forgotten that they are not used to money. Wishing to placate the first one, I gave him three 3-rouble notes and one 1-rouble, instead of the single 10-rouble note. Now the other guide was angry, because he had only half as many bits of paper.

To restore peace it was necessary to give them exactly the same. You should have seen how pleased they were when they set off for their homes.

After the twenty-four hours rest, everyone was in good spirits - cheerful and happy. That day we covered about 18 kilometres, and pitched camp by the River Kate-Taban. Here there were three yurts[2] and one fanza,[3] the latter called Sidungou.[4] It was occupied by two old men - one a *taza*[5] and the other a Chinese sable[6] trapper. They proved to be very hospitable, and did their best to satisfy our needs.

I particularly wanted to climb up to the Khorski Pass and so inquired about the route. The *taza*, whose name was Kitenbu, agreed to guide us. He was probably about sixty years old; he already had some silver hairs, and his face was very wrinkled. In appearance he could have been taken for a Chinese, and I knew him to be a *taza* merely because he said so. He told us that he formerly lived by the River Ussuri, but pressure from Russian settlers forced him to move to the River Bikin; his home had been there for the last ten years.

Kitenbu immediately started to get ready. He took a patched blanket, a goatskin, and an old Berdan rifle that had seen better days. I took a kettle, notebook and fur sleeping bag, while Dersu had his tent, pipe and the provisions. Accompanying us were my dog Alpa, and Kitenbu's Kady - with his pointed muzzle, erect ears and grey fur.

The Kate-Taban was a substantial stream. Its valley was narrow low down but broadened out higher up. The mountains around were covered in old mixed conifer-broadleaf forest.

In the morning the weather was fine. We expected that by evening we would reach a hunter's cabin on the far side of the watershed. However, our plan was not destined to work out. In the afternoon the sky clouded over, rings[7] appeared around the sun, and the wind got up. I was already thinking of turning back, but Dersu assured me that there would not be a blizzard: the worst would be strong wind, which would die down the next day. This is in fact how things turned out. At four o'clock the sun disappeared, and we could not tell whether it was in cloud or fog. The air was filled with dry powdery snow.

The wind rose and cut our faces like a knife. When darkness fell we were on the ridge itself. Dersu stopped, and started talking with the old *taza*. Approaching them I gathered that Kitenbu had strayed from the path somewhat. To avoid getting lost they decided to spend the night in the open.

'Captain,' Dersu turned to me, 'we no find hut today, must sleep here.'

'Very well,' I agreed, 'let's find a good spot.'

My companions continued deliberating, then moved away about twenty paces and began taking off their packs. It could hardly be called an ideal place to bivouac. It was on a level saddle covered in dense forest. Off to one side stretched a long spur, terminating in a low conical hill. On both slopes of the pass were thickets of pine scrub, with some other bushes covered in dead dry leaves. We made our way into the most sheltered thicket, and settled down at the foot of a massive pine about twenty metres high.

Dersu went off with his axe for firewood while Kitenbu began cutting conifer branches for bedding. Meanwhile I got a fire going. It was half past six before we had everything ready, by which time we were exhausted. Once the fire was blazing away the camp seemed more cheerful. Now we could get our boots off, dry our clothes and think about supper. In half an hour we were drinking tea and discussing the weather.

My Alpa's coat was not as thick and warm as Kady's. He was shivering and, tired out by the journey, was lying by the fire; his eyes were half shut, and he seemed to be dreaming. Kady, most of his life used to hardship, paid little attention to the discomforts of a nomadic life. Curling up he immediately fell fast asleep, and was soon covered in a dusting of snow. Occasionally he got up to stretch, then walked on the spot for a few moments, lay down on his other side, and poked his muzzle into his chest to warm himself with his own breath.

Dersu always felt sorry for Alpa. Before taking off his boots, he would first make the dog a bed of spruce branches and dry herbage. If neither was available, he would give the dog his jacket.

Alpa understood Dersu's kindness. During stops for rest Alpa would seek out Dersu, prance around him, paw him and in every way possible try to attract his attention. As soon as Dersu picked up his axe, Alpa would calm down and wait patiently until he came back with an armful of spruce branches.

We ourselves were as tired as the dogs, and so as soon as we had drunk our tea and heaped more wood on the fire we prepared to turn in.

We each found our own position around the fire: I chose a spot downwind from it, while Dersu settled down off to one side. He rigged up a sort of tent with his sheet, and threw a coat over his shoulders. Kitenbu sat down at the foot of the pine and covered himself with his blanket. He agreed to keep watch and look after the fire all night. I cut some spruce branches for a mattress, spread my sleeping bag on them, and made myself comfortable between a fallen tree and the fire.

A storm in high forest is always terrifying. You imagine that the tree you are sleeping under will fall and crush you. So, despite being so tired, for a long time I could not get to sleep. Some demon was whipping the wind up into such a frenzy that it was like a savage beast throwing itself at everything in its way. The trees in particular caught its fury: it was a pitched battle between the forest giants and the unbridled elements. The wind came in roaring gusts, then rushed away and mournfully howled in the distance. I had the impression of being in the very heart of a gigantic whirlwind. The wind blew in a huge circle, returning to our camp and attacking once more the pine, doing everything in its power to uproot the tree. Against the mighty assault the giant tree stood firm, sullenly swaying from side to side. I recalled a storm near Lake Khanka,[8] and a blizzard when crossing the Sikhote-Alin mountains. I heard the old *taza* putting more wood on the fire, and a roar as the wind fanned the flames. Then everything became confused as I nodded off.

About midnight I woke up. Dersu and Kitenbu were already awake, and were talking. There was a note of anxiety in their voices. The thought crossed my mind that the pine might come down. I quickly threw back the hood of my sleeping bag and asked what was up.

'Nothing, Captain, nothing at all,' answered Dersu. But I sensed that he was not telling the truth: he simply did not want to alarm me.

The fire was burning brightly. Dersu was squatting by it, shielding his face from the heat, adjusting the logs and gathering the embers together. Kitenbu was stroking his dog. Alpa was sitting beside me, shivering.

It was as if all the evil spirits had assembled and were tearing about in the taiga one after the other, howling and crying, trying to wreck the world and take nature back to a state of chaos. Our ears were assaulted by frenzied cries and groans, wild laughter and howling. Suddenly all would go quiet, and only then could we make out what was happening. The brief spells of calm told us that the wind would soon die down.

The fire was blazing. Black shadows and splashes of red light chased each other, jumping around over the bushes and banks of snow. They would alternately recede from the fire, then race back to it.

'Nothing, Captain,' Dersu repeated. 'You sleep. We just talk.'

I did not inquire further, pulled up the hood, and went back to sleep. About half an hour later I was woken up again by voices.

'Something wrong', I thought to myself, and crawled out of my sleeping bag.

The storm was dying down a little. Here and there stars could be seen. Each gust of wind swept dry snow along the ground with a noisc like sand. I saw my companions by the fire. Kitenbu was on his feet and listening intently. Dersu was standing sideways on, shielding his eyes from the light of the fire, and staring into the dark. The dogs were also alert; they would get close to the fire, apparently to lie down, but would then jump up and move away somewhere else. They could sense something, and kept looking in the same direction as Dersu and Kitenbu.

The wind made the fire roar, throwing up thousands of sparks, swirling them around, and carrying them off into the depths of the forest.

'What's going on?' I asked Dersu.

'Pig on the move,' came the answer.

'So what?' It seemed quite natural to me: the wild pigs[9] were moving through the forest, came across us, and now were noisily expressing their annoyance.

Dersu gestured impatiently, and said, 'How you be in taiga so long, and understand nothing? In winter pigs no move around at night.'

From the direction the two men were facing could be heard the snapping of twigs and the characteristic sneezing of the pigs. The animals came down from the watershed, passed close by us, and continued along the side of the conical hill. I stretched my legs, and did not feel like sleeping any more.

'So why are the pigs moving at night?' I asked Dersu.

'They no move if no reason,' he replied. 'Somebody chase them.'

I thought he was referring to some Udege, and was wondering how they could be moving around in the forest on skis at night. Then I remembered that, for Dersu, not just humans were 'people'. Suddenly it dawned on me: the pigs were being pursued by a tiger - and it must be very close to us.

Order began to take over again from the recent chaos in our surroundings. Some other force was getting the upper hand over the wind, and calming it down. But, judging by the continued swaying of the pines, it was clear that high overhead the battle was still raging.

Without bothering to brew tea I pulled my sleeping bag closer to the fire, crawled in and went back to sleep. It felt as though I slept for a long time. Suddenly a heavy object struck me in the chest, and at the same moment I heard a dog's yelp and Dersu desperately shouting, 'Quick!'

I immediately threw back the outer flap of the sleeping bag. Snow and dry leaves hit me in the face, and at that very moment I saw a long shadow gliding past among the trees. On my chest lay Alpa.

The fire was almost out: just two logs were smouldering. The wind fanned the embers and blew sparks over the snow. Dersu was sitting on the ground, with his legs buried in the snow. He held his chest with his left hand - as if he were trying to stop his heart beating. The old *taza* was lying face down, not stirring.

For a few moments I could not imagine what had happened, nor what I should do. With some difficulty I managed to push Alpa away, then crawled out of the bag and went over to Dersu.

'What has happened?' I asked him, shaking him by the shoulder.

'Amba, Amba!' he shouted, terrified. 'Right through camp! Take one dog!'

Only then did I notice that Kady was not with us.

Dersu rose to his feet and began building up the fire. As soon as it was giving out some heat, Kitenbu came to. He glanced around with fear written on his face, looking as if he had gone mad. In any other situation it would have been comical.

This time it was I who kept most self control. It was because I had been asleep and had not seen what happened. Soon, however, our roles were exchanged. As Dersu calmed down, I became more anxious: what was to stop the tiger returning and attacking us? How had it all happened, and why had no-one fired?

It turned out that Dersu was the first to awake - on account of the dogs. They were jumping about from one side of the fire to the other. To save himself from the tiger, Alpa jumped straight at Dersu's head. Still only half awake, he pushed the dog away and at the same moment saw the tiger close by. The terrible beast seized the other dog and slowly, as if knowing that nothing could stop it, carried him off into the forest. Frightened by Dersu's shove, Alpa jumped over the fire and landed on my chest. Then I heard Dersu's shout.

Instinctively I reached for my rifle, but had no idea in which direction to fire.

Suddenly there was a rustling in the undergrowth behind me.

'There!' whispered Kitenbu, pointing to the right of the pine.

'No, there!' answered Dersu, pointing in the opposite direction.

There was more rustling, this time from both directions. The wind in the trees made hearing difficult. Now and then I thought I could indeed hear twigs snapping and that I even saw the tiger itself, but soon realised that I was mistaken: it was just a log, or a young spruce. The undergrowth was so thick that even in daylight little could be seen through it.

'Dersu,' I said, 'climb a tree; you'll be able to see better.'

'No,' he replied. 'No can. Me too old. Me no more climb trees.'

Kitenbu also refused to climb. So I decided to try myself. The trunk was smooth, and plastered with snow on the windward side. It was hard work climbing just three metres up the pine. My hands were soon frozen and I had to come back down.

'No need,' said Dersu, looking at the sky. 'Soon dawn.'

He picked up his rifle and fired into the air. Just at that moment there was a gust of wind, and the report was drowned in the commotion.

We made a big fire and brewed tea. All the while Alpa pressed up close to me or to Dersu, and at the slightest sound flinched and looked around in fear.

For about forty minutes we sat around the fire and compared our impressions.

At last, light appeared in the sky. Soon indistinct shapes became visible, and the stars began to go out, just as if they were speeding away into the depths of the universe. A short while later the blood-red dawn lit up the eastern sky. The wind died away rapidly and the frost became more severe. Then Dersu and Kitenbu went over to the bushes. The spoor told them that nine pigs had passed by, and that the tiger was large and old. It spent a long time circling us, and attacked the dog only when the fire had gone out.

I suggested to Dersu that we leave our things in camp and follow the tiger's track. Expecting him to reject the idea, I was surprised when he agreed that we should recover Kady's body.

He began explaining that the taiga gives plenty of food to the tiger, and it is forbidden to attack people. This tiger had been tracking the pigs, but saw us, attacked our camp, and took the dog.

'Such Amba we can shoot, no sin,' he concluded at last. After a quick breakfast of cold meat and hot tea, we strapped on our skis and set off to track the tiger.

The weather was once more calm. The ancient spruces and pines had shed their loads of snow, leaving great piles of it scattered about. They were lit up by bright sunshine, so that the forest took on a festive air.

The tiger had gone back along its own track, which led us to a fallen tree. The spoor went right under it.

'No hurry, Captain,' said Dersu. 'No need go straight, go round, look well.'

We made our way around the obstacle.

'This way!' called Dersu suddenly, and set off in a new direction.

Here it could be seen that the tiger had sat for a long time in one spot. The snow under it had melted. It had dropped the dog in front of it, and listened in case anyone was giving chase. Then it went on.

We continued tracking for three hours. The tiger did not follow a straight line, but chose areas with little snow, thicker vegetation and more fallen trees. At one place it had climbed up onto a fallen tree and stood there for a long time, but was suddenly startled by something, jumped to the ground, and crawled on its belly for several metres. Now and then it stopped and listened. On our approach it set off, at first with a few leaps, and then walking and trotting.

At last Dersu stopped and began to consult with Kitenbu. He advised turning back. His reasons were the tiger was not wounded, there was not enough snow, and to carry on would be a waste of time.

I was wondering why the tiger had not eaten the dog, and was carrying it so far. As if reading my thoughts, Dersu remarked that it was not a male tiger but a tigress, and she was taking the prey to feed her cubs. She would not take us to her lair, but would instead lead us all over the place until we got tired. It was a watertight argument.

When we were all agreed, Dersu faced the way the tiger had gone and shouted, 'Amba! You worth nothing. You robber, worse than dog. Me no fear you. Next time me see you, me shoot.'

Then he lit his pipe and set off back along our ski track.

Shortly before arriving at our bivouac, it happened that I was ahead of the others. When I reached the top of the pass, I fancied I saw something disappearing down the slope from our camp.

Indeed, all our things were scattered around and ripped up. Of my sleeping bag just a few pieces remained. Tracks in the snow showed that the culprits were two wolverines.[10] It was they I had seen.

We salvaged what we could, then quickly descended the mountain side and went back to rejoin the soldiers. It was easy going, since our ski track was still frozen though dusted with snow. We flew down into the valley, and by evening were once more with our comrades.

Chapter 5

NATIVE PEOPLES, SETTLERS AND TIGERS

The massive long-furred Amur tiger still prowls the rich temperate mixed forests of the southern Russian Far East, Northeast China (previously known as Manchuria) and North Korea. In this region the great striped cat has been, for tens of thousands of years, the top predator and an indicator of the health of a remarkable plant and animal community.

For thousands of years there has also been a human presence in the forests of northeast Asia. To the aboriginal Manchu-Tungus people tigers were masters of the taiga and played a powerful symbolic role in their myths, legends and art.[1] Some of the first representations of tigers take the form of stone carvings along the banks of the Amur River, dating from 4,000-3,000 BC. In the Amur-Ussuri region the Nanay, Udege,[2] Orochi and Ulchi peoples revered Amba, the tiger, as a totemic animal, sacred ancestor, guardian of the forest and protector of ginseng. Indeed the Udege were often called 'Tiger People'. Native peoples were so much in awe of the great striped cat that they did not even dare mention its name. If they came upon its tracks in the forest they would bow respectfully and quietly retreat. They would never kill tigers except in self-defence, though these attitudes changed with more intensive hunting by the Chinese and Russians in the region in the later nineteenth century.[3]

These aboriginal tribes are probably the descendents of once powerful Manchu peoples such as the Bohai and Jurchens.[4] The Udege, a small Tungus tribe, made their home for centuries in the fastnesses of the Sikhote Alin.[5] They lived in family groups that followed a nomadic hunting and fishing way of life in the vast wilderness, dwelling in log or birch-bark huts. In summer they transported loads or their whole families by river in boats hollowed out from a poplar tree with an axe. These boats were often decorated with carved tiger motifs. In winter they travelled on long skis and used sledges pulled by dogs or by themselves.

The Chinese called the Udege and related tribes 'Fish-skin Tatars'. The Udege men wore embroidered *izyubr* or fish-skin jackets, leggings, boots and caps with a hunter's veil or mantle extending over the neck and shoulders as a protection against insects, sun and snow. The women wore more elaborately embroidered robes, decorated with patterns of spirals and coils indicating Chinese influence. Udege hunters showed extraordinary skill, stamina and resourcefulness in a harsh environment, even sleeping on a bed of grass in the snow during the winter: Arsenev described them as true 'People of the Forest'. They hunted elk throughout the year for their meat - which is rather like coarse dry beef - and for their hides, using spears, bows and arrows, cross bows and later guns. The deer called *izyubr*[6] were hunted especially during the autumn rut, lured with the aid of birch bark pipes. Musk deer, killed for their valuable musk as well as meat, were also lured or trapped with snares.

It has been suggested that tigers may also call up *izyubr* by imitating their calls, though as tigers make similar sounds at other times there is no definite proof of this. In Arsenev's *Dersu Uzala*, he and Dersu were on one occasion observing the *izyubr* rut, and 'suddenly Dersu stopped and began listening intently. He turned back and then froze. The roar of an old stag could be heard, except that it did not sound quite right.

'"Hm, you know what he is?" Dersu asked quietly.

'I said I thought it was a stag, but an old one.

'"He tiger," he whispered. "He very cunning. He always deceive stag like this. Stag not know yet who is roaring. Tiger will soon take a hind."

'As if to confirm what Dersu had said, the stag answered back with a mighty roar. Immediately the tiger again answered the stag. It was a good imitation, except that at the end came a short purr.' [7]

The Udege and related tribes believed in the masters of nature - especially the Master Spirit of the taiga, and the spirits of the mountains, the forest and individual animals who were totems. One of the masters of nature especially revered was the 'great sovereign', Amba, the tiger - regarded by the Udege as well as by the Nanay and Orochi as a totem and sacred animal.

Arsenev recounts a typical story from his time. 'This year on the River Anyuy there were many tigers. They were behaving audaciously, coming up close to houses. About a fortnight before our arrival one tiger showed particular daring: almost every day it was visiting the Udege, often in broad daylight. At first they persuaded it to leave them in peace, but then had to threaten it with spears and guns. Eventually they conducted shamanistic rituals at night, and begged *Buyn Atszani* - the god of the animals - to call off "his hound". But to no avail. The tiger wandered along the other bank of the river, lay down on the ice opposite the Udege yurt, yawned, and generally scared the people. Then the Udege realised that the terrible beast had of its own free will left its owner, *Buyn Atszani*, that it was no longer subject to Him, and therefore that killing it would not be a sin.

'One evening a woman wanted to leave the yurt to fetch water. The moon was full. No sooner had she opened the door a crack than she saw the striped monster close by. She quickly slammed the door to. Then one of the Udege men cut a hole in the bark wall of the yurt, and through it fired at the tiger. It dropped, but then managed to crawl away and along the ice on the river. In the morning they found it dead on the river ice near the far bank. The Udege men covered it in snow, and alongside set up two stakes with streamers, to make it known to any newcomers that it was a forbidden place.'

Arsenev wanted to collect the remains, but the Udege were opposed to the idea: if he did, they would not be left in peace by the other tigers. Even so, he went back after a couple of days to take some bones, but was disappointed to find that they had all gone. At the spot was a hole in the ice, with the river flowing swiftly past under the remaining ice. This convinced the Udege that *Buyn Atszani* did not allow blasphemy against his 'dog', even though it had taken its freedom.[8]

Shamans were special people regarded as being in contact with the spirit world. They practised a hunters' religion, with rituals designed to ensure success in the hunt; they also prayed for the souls of animals needed for food, healed the sick, and sent the souls of the dead to the afterlife.

According to Udege mythology the sun was a woman with whom the man in the moon was in love, and a female shaman would carry out the ceremony of the setting sun. Other worlds were believed to exist beyond the sky and below the earth, and the earth itself was seen as a living being - a recumbent dragon below which there was another, like a mirror image, in the sea.[9]

During the 18th and early 19th centuries there was increasing contact between the native people and the Chinese. The Ussuri region was part of the Chinese Empire until annexed by the Russians following the Treaty of Beijing in 1860, and even before this Chinese were already settled in these mysterious forests. The native tribes were subjected to the extortion and sadistic cruelties of the *honghutzi* Chinese bandits.[10]

They were also exploited by ordinary Chinese traders who sent agents into the territory to set up trading posts, dividing the region for their operations into districts containing both the aboriginal people and Chinese settlers. The natives had to exchange large quantities of sable furs, ginseng and deer antlers for Chinese padded coats, iron tools, guns, rice flour, tea and tobacco, but especially opium and alcohol. Reduced to indebtedness and dependence on the Chinese traders, many Udege were forced to mortgage the following season's hunting catch in advance or even to surrender their women in lieu of debt.[11] Many Udege adopted Chinese ways including the language, and as most of the Chinese immigrants were single men intermarriage resulted in a mixed population in the southern Ussuri region, who were known as *tazy*.

With the Bolshevik Revolution of 1917, efforts were made to send Communist teams into these remote regions to make contact with the tribes. The Udege assisted the Red Partisans against the anti-communist forces during the Russian Civil War, as described in A. Fadeyev's novel *Last of the Udeges*. At first the arrival of Soviet power had limited influence in this wild country, and the native people were treated with some consideration by the Committee of the North,[12] but with collectivisation in the 1930s the tribes were forced to give up their nomadic ways and to settle down. Collectives were organised for hunting, fishing and gathering medicinal plants. There was also diversification into other activities such as forestry, vegetable growing, collecting cork from the Amur cork trees, and beekeeping. As shamans and traditional beliefs were suppressed, many aspects of Udege culture seemed doomed, and even the distinctiveness of the people themselves was threatened by intermarriage with Russians.

With the end of Soviet power attention was paid in the 1990s to reviving the traditional culture and language of the Udege. But there were also negative results from the fall of Communism. After the collapse of the state fur industry, there was domestic economic depression and reduced

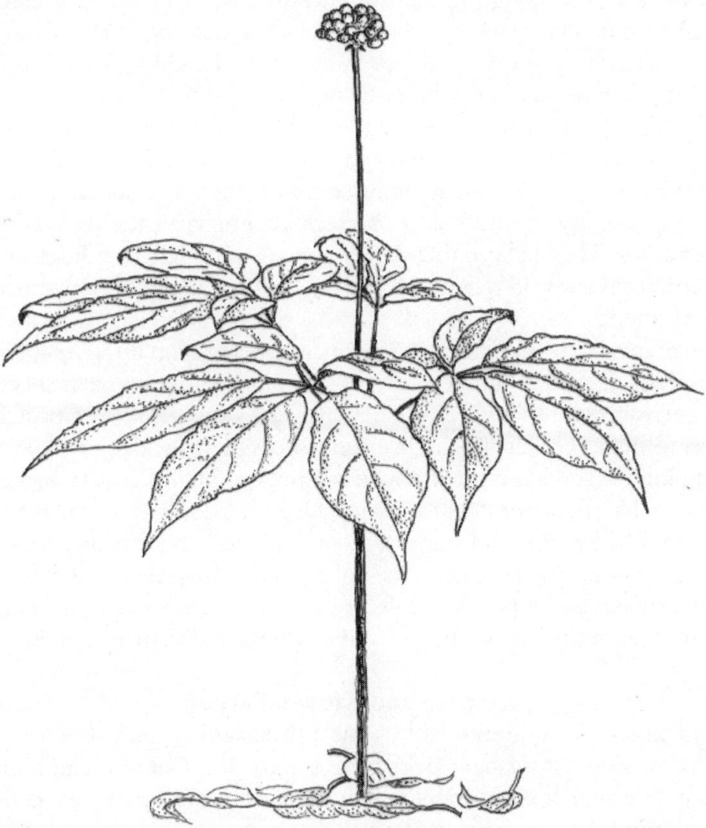

Ginseng is an unmistakeable woodland herb . At the top of the main stem springs not only a thin stalk carrying the flowers, or later red berries, but also the four leaves in a whorl, each divided into five leafletes. Together they form a flat 'table' to catch the light. In contrast to the slender aerial parts, the medicinally valuable part of the plant is the bulky root underground.

international demand, and so the Sikhote-Alin began to attract traders in search of tiger pelts and bones, bear gall bladders and wild ginseng. Some Udege have been attracted by lucrative offers for these products.

Today, about half of the Udege population of 2000 lives in the village of Krasny Yar in the middle reaches of the Bikin Valley, and others in the Sukpay, Khor and Samarga areas farther north. After the end of Soviet power foreign logging companies invaded Primore. Clear felling of timber in the Bikin catchment area by the Hyundai Corporation in the early 1990s threatened the forest homeland of the Udege and in 1995 they formed a Druzhina or community defence force - an anti-poaching team operating in the Bikin Valley as part of Inspection Tiger.[13] Many Udege supported this community struggle against poachers, loggers and the lure of foreign money, in an effort to maintain their cultural integrity and subsistence livelihood, including hunting.[14]

At that time some fifty Udege hunters in the Bikin Valley rotated hunting plots so as not to diminish wildlife populations. After an area is hunted it is left alone as a 'quiet zone' for a period. The Udege consider the presence of tigers as an indication of the well being of the forest. Although tigers and humans hunt some of the same animals, such as wild pig and *izyubr*, the Udege hunters do not view the tiger as their competitor. They believe that poaching, deforestation and the resulting loss of food for wild pigs and other animals is a much greater threat to their hunting.

In many parts of Russia the development of natural resources, including oil, gas and timber, continues rapidly with minimal respect for the environment. In the face of such pressures, the creation of protected areas to preserve valuable habitats and ecosystems, endangered wildlife populations and traditional native cultures is vital.[15] In the southern Russian Far East the inclusion of Udege and Nanay tribal lands as protected areas plays an important part in the habitat protection plan for the Amur tiger. Threats to the tiger and its ecosystem, whether these be illegal logging or poaching, also damage the livelihood and culture of the indigenous peoples who share the forests and their resources.

Chinese and Korean Settlers

The Chinese were present for about a thousand years in Primore and Amur Region. Amur tiger pelts were sent to the Chinese emperors as tribute from at least the time when Tang Dynasty forces invaded the area in the 7th century, but before the 19th century there were few human inhabitants in the Amur tigers' domain. Tigers were abundant in a virtually untouched wilderness, like a vast natural reserve - the *shou hai*, or forest sea. Both the native tribes and the Chinese set up shrines in the remote forests to Da Sheng the great spirit of the 'Lord Tiger'.

Arsenev came across several of these shrines. 'On the pass itself, on the right side of the path, stood a small shrine made of planks. Inside there was a cheap print depicting Chinese gods, and in front of it lay two wooden boxes with the remains of paper tapers. On one side lay several tobacco leaves and two lumps of sugar: these were an offering to the god of the forest. On a nearby tree was hanging a red piece of cloth with an inscription which can be translated as "For the Lord, the true spirit of the mountains. Long ago in the State of Tsi he was commander-in-chief of the Da-Tsin Dynasty, and now protects the forests and mountains".' And at another shrine the inscription was 'To the Prince [ie the tiger] that dwells in the forests and mountains. In olden times, in the Han Dynasty, he saved the State. Now he is the spirit giving happiness to people'.[16]

The number of Amur tigers killed by aboriginal hunters and by the Chinese settlers was at first limited. The population of the big cats probably remained unchanged until the 1870s and 1880s when tiger pelts and body parts became a source of profit for increased numbers of Chinese and, later, Russian settlers and traders in the Amur-Ussuri region. So began the destruction of the Amur tiger population.

However, Amur tigers had long been hunted for their fur. Crossing the sea in junks the Manchus took tiger and leopard skins, sable pelts and wild ginseng to trade in Japan for silks and brocades. It was noted that the 'Manchurian tiger' was not only valued for its splendid pelt but also for its 'whole carcase, for the Chinese believe that the bones, blood, flesh and heart of the tiger have medicinal properties of rare power and will pay a goodly price for decoctions brewed by the apothecary that contain such ingredients as powdered tigers' kneecaps and tigers' blood'.[17] Tiger bone was and is regarded by the Chinese as especially efficacious in the treatment of rheumatism, and tigers' blood was used for debility. The Chinese trade in tiger derivatives for traditional medicine was a major factor in bringing the Amur tiger close to extinction in the Russian Far East, Northeast China and Korea.

In the Amur-Ussuri region the first Chinese came seeking ginseng and hunting leopards and tigers. They were soon followed by merchants, and then by settlers who worked the land in their own diligent and successful way when the Russians took over the Primore Territory in 1860. After 1870 the Qing authorities eased restrictions on Chinese migration into their Manchu homeland and many famine-stricken peasants, especially from Shandong Province, settled in Manchuria and the Russian Far East. Despite the desire of the Tsarist authorities to enhance Russian predominance there was still a need for Chinese labour and food producers in these thinly settled regions. Chinese and Korean peasant farmers provided most of the fresh vegetables for Vladivostok,

Khabarovsk and other towns in the Russian Far East. Chinese labourers, many of them seasonal workers, provided most of the labour on the Chinese Eastern Railway and the Amur and Ussuri sections of the Trans-Siberian Railway built in the 1890s.

There were still some 24,000 Chinese in Vladivostok in the 1920s, with their own schools, theatre and newspapers. These were closed by Stalin in 1937-1938, and the Chinese were expelled and sent to labour camps where most of them perished.[18] A thousand-year-old Chinese presence in the Amur-Ussuri region seemed at an end. In the 1990s, however, with the collapse of Soviet power, the borders were once again open to Chinese traders and workers who rented plots of land for cultivation from the local authorities. The growing numbers of Chinese traders dealing in wildlife products including tiger parts once again threatened the great striped cat.

Koreans settled in the Russian Far East have also had an impact on the wildlife, including the tiger. Long before the Russians arrived there were Korean villages in the frontier area around Posyet. Dressed in their distinctive white flowing robes, the Korean settlers were skilful cultivators of millet, beans and rice, and they also reared pigs and cattle. Being hardworking and enterprising they were highly regarded, and were encouraged to settle with grants of land by the Russian authorities.

In 1937, because of Soviet tensions with Japanese power in Manchukuo (Japanese-occupied Manchuria), nearly half of the Korean population was deported from the frontier areas of the Russian Far East to Kazakstan. A larger number remained in the Russian Far East, however, and much later Soviet Koreans, who had maintained contacts with relatives in Korea, played a significant part in the illicit trade in wildlife products, including tiger parts.

In the Chinese and Korean cultures tigers were associated with strength and power. The tiger was regarded as the King of Beasts, and tiger motifs signified the power of kings and emperors. In Korea the White Tiger was revered as the mountain spirit, the guardian of the west, while the Blue Dragon was the guardian of the east. In one Korean legend, a beneficent tiger rolled back the boulder blocking the tomb of a respected old man so that his spirit would be freed.[19]

In China, the markings on a tiger's forehead are interpreted as *Wang*, or king, and this had special significance for the Chinese because of their reverence for calligraphy (see diagram on the next page). 'But add to this its enormous strength and ferocity, its cruel cunning, and the wild and lonely mountain fastnesses and gloomy forests it chooses for its lair, and it will be seen how readily the tiger lends itself to a prominent role in myth and legend in a country where these hold as great a place as they do in China'.[20]

Daoism emphasises a reverence for nature and the need for human beings to be in harmony with natural forces. Tigers and dragons together symbolise the two great forces of nature - *yin*, dark, mysterious or evil, and *yang*, bright or good.

In Buddhist thought the role of the tiger is reversed, and it represents *yang*. However, just as dark and light are both necessary, in Chinese mythology the breath of the tiger creates the wind, the breath of the dragon - the clouds, and together they produce the rain that regenerates the earth.

Reverence for the tiger in Chinese culture as a symbol of strength, courage and kingly power did not mean, however, that the animal itself escaped persecution. Chinese civilisation regarded the tiger with veneration but has at the same time also brought it to the verge of extinction.

Arsenev argued that as the Chinese had already devastated much of the wildlife in their own country they then turned their attention to the Russian Far East in the 1900s.[21] The market for tiger derivatives for traditional Chinese medicine, in the hands of Chinese traders, was one of the main causes of the decline in numbers of the great cat.

The heads of three Amur tigers, to show the variation in patterns of marks on the face. Particularly useful in distinguishing individuals are the marks above and below the eyes. The pattern on top of the head also varies, and occasionally resembles the Chinese 'wang'

Russians and Tigers (1860-1939)

The exploitation of wildlife, including tigers, was one of the reasons why the Russians came to the Amur-Ussuri region and the Pacific seaboard of the Russian Far East. After they founded Vladivostok in 1860 it was not unusual for the tracks of the great cat to be found in the snow in outlying streets of the town. Their presence is commemorated in street names - Tiger Street, Tiger Hill, and so on - and the great cat is still the centre piece of the city's municipal crest. In the late 19th century the citizens often put up barriers around their houses, as some of the tigers were man-eaters. Domestic dogs were often preyed upon,[22] even within the town, and in many places in the region settlers found it almost impossible to rear horses and other domestic livestock because of the great cat's depredations.

The hunting of Amur tigers - for their fine pelts, trophies, and body parts to be used in traditional Chinese medicine - started to have a serious effect on their numbers in the 1870s and 1880s. The original population of the Amur tiger in the Russian Far East has been estimated to have been around 1,000 animals.[23] It was not much reduced until the late 19th century, when the destruction of the tiger population began in earnest as increased settlement brought more hunters into the forests. Tigers were regarded as 'harmful' animals by the government, and so could be hunted without restriction. Although no enumeration of the population of the great striped cats was made at this time, it was recorded that as many as 150 were shot or poisoned each year in the 1870s and 1880s,[24] and 50-60 a year in 1900-05.

In the late 1890s and the early 1900s, the Russian Far East and Manchuria were opened up through the building of the Trans-Siberian and Chinese Eastern railways. The policy of the Tsarist government in encouraging settlement of the Russian Far East and extending Russian influence in Manchuria had a considerable impact on the wildlife of the region. Sable fur, deer antlers 'in velvet' and ginseng, as well as tiger and leopard pelts and bones, were all eagerly sought after.

Russian hunters preferred to track the tiger down in the snow during the winter, when its trail could be followed and when its fur was in prime condition. These intrepid tiger hunters usually worked in teams of four: two would

The crest of Vladivostok

follow the trail while the other pair would transport equipment, pitch camp and prepare a hot meal for their comrades; later, the two pairs would change over. Once fresh tracks were found the tiger would be pursued with dogs, often for a week or more, before it was brought to bay and shot. Tigers were also killed with gun traps - rifles attached to trip wires - and poison.[25] Nikolay Baykov left accounts of these tough tiger hunters, who were themselves really 'wild men of the woods'.[26]

In addition to the killing of tigers for their pelts and for medicine, there were added pressures on the tiger population. These included live capture for zoos, since a live animal was worth more than a skin. Capture teams worked in a similar way to hunters, pursuing the animals with dogs. Young animals up to two years old were separated from their mothers, cornered, pinned down with forked sticks and then tied up. Sometimes the mother would be killed defending her cubs. The capture itself was a great shock to the young tigers and not all survived. Earlier in this century Hagenbeck of Hamburg would pay up to £1,000 for a live Amur tiger.

More Amur tigers were captured between 1925 and 1947 than were killed by hunters, and live capture for zoos was a major reason for the reduction of the wild Amur tiger population.[27]

The completion of the railways opened up the Russian Pacific seaboard and Manchuria to settlement, especially along the line from Khabarovsk to Vladivostok. Towns and villages grew up and there was widespread deforestation with uncontrolled felling and burning as land was cleared for agriculture. This was especially true of the Khanka lowlands, where the reed beds had been prime tiger habitat. As well as cattle rearing, wheat and vegetables, Chinese crops such as millet, rice and soya beans were supplemented by melons and sugar beet brought in by Ukrainian settlers.[28] In the years before the First World War, as settlement and cultivation spread along the main river valleys, some of the best lowland habitats of the Amur tigers were destroyed and there was increasing tiger-human conflict.

Hunting prior to the First World War greatly reduced the number of the tigers' prey species such as wild pig, *izyubr*, sika and roe deer, and goral. Tigers were also affected by natural calamities. In the 1880s, according to Arsenev, many tigers migrated from the Sungari River region of Manchuria eastwards into the Sikhote-Alin as widespread disease had reduced the numbers of their prey animals.[29]

Similarly, as a result of the exceptionally severe winter of 1914-1915, there was a reduction in the number of hoofed animals as shortage of fodder prevented wild pigs and deer from reproducing. By the 1920s Amur tigers survived in only about five isolated localities in the Russian Far East.[30] There were also three successive severe winters between 1924 and 1927, with heavy snowfalls that made movement and finding food difficult for both predators and prey.

The Japanese intervention and occupation during the Civil War (1918-1922) involved the pillage of timber and other natural resources from the Russian Far East on a large scale. The number of Amur tigers continued to decline in the 1920s and 1930s. In the 1934 hunting season more than twenty tigers were killed in Primore, whereas the usual number taken was only ten a year.[31] During these years there was no protection for wildlife in the chaos that prevailed in Manchuria, torn apart by warlords and Japanese intervention in the 1920s and 1930s.

In the 1930s, despite their rarity, tigers could still be hunted in the Russian Far East throughout the year. They were given some protection with the setting up of the great Sikhote-Alin and Lazovski reserves in 1935 and 1937, and also benefited from temporary restrictions on the harvest of *izyubr*. However, hunting and capture of tigers continued even within the reserves. A number of tigers were killed in order to secure museum specimens; Moscow Zoological Museum has a large collection of Amur tiger pelts and skulls.

Soviet economic development in the 1930s at the time of the early Five Year Plans - logging, pulp and paper mills, mining, the building of power stations, and the extension of cultivation - continued to destroy tiger habitat. There were years of winter starvation for wildlife with heavy snowfalls in 1932-33, 1935-36 and 1938-1939, though Evgeni Matyushkin believed that Amur tigers can cope with hard winters and that these do not significantly affect their overall status.[32] Even so, despite the extent of the virtually untouched Sikhote-Alin forests, hunting meant that very few Amur tigers survived by the late 1930s, and concerned zapovednik scientists - including Kaplanov - were calling for their protection.

Chapter 6

MAKHA AND THE WOUNDED TIGER[1]

The Udege were speaking the truth. With so many tiger tracks here, one might almost imagine that, however many there were in Ussuri Region, they all came to the Anyuy for the winter. However, I soon concluded that in fact there were only two - an old track of a large animal, and a fresh track of a younger one. The tracks went along dry streambeds in both directions, crossed from one side to the other, formed circles and went back on themselves. In one place we came across very fresh spoor. Although Krylov unleashed the dog, it unexpectedly did not race ahead but, burying its nose in the snow, began sniffing every bit of herbage we passed. After a minute or so Kady once more set off along the channels, but only very cautiously, then stopped and listened. Twice it lay down, with its hind legs tucked under it and its front ones extended.

From its behaviour and alert ears it was clear that it sensed something near. The dog's nervousness communicated itself to us too: I was not confident that, if a tiger had that minute lept up, I could have fired calmly. Kady refused to go on, and we too remained rooted to the spot in agonising suspense. At last Krylov could stand it no longer. He approached Kady, stroked it, and looked around. Kady did not stir. Then we both stepped over it, and Kady quietly got to its feet and followed us. After about ten paces Krylov sent Kady on in front, but after several hundred paces it again stopped. Pricking its ears it began looking around, then suddenly ran back. Dashing past Krylov it pressed itself against my legs. I began stroking it, and could feel it shaking.

It meant that the tiger was close by.

I was filled with curiosity and fear. My heart was pounding, and every beat seemed to be audible. My breathing was also affected. We both stood stock still, straining our ears and eyes, but not a single rustle disturbed the peace of the forest. Suddenly a frozen tree cracked, followed by an echo. It felt as though an electric shock had ripped through me from head to toe. Krylov immediately turned towards the source of the sound; Kady flinched and kept even closer to us. It was some time before we realised it was just a trick of the frost. Now Kady once more advanced. We followed. Its behaviour was most odd - walking in a curve, staring into the centre of the space it was avoiding. It made us alert. In the centre of the space was an enormous leaning poplar, almost lying along the ground. Near the tree the snow was trampled, and elsewhere it was churned up. Something could be seen under the tree. For some time we remained standing in a state of indecision. Spotting something by my feet, I looked down and picked up a frozen piece of fern root. Krylov took it and threw it towards the poplar; but nothing stirred, so we approached the tree more boldly.

It turned out to be a tiger's lair. The poplar was hollow from one end to the other, with the opening on the underside. The ground under it was free of snow, and the dry herbage was well flattened. All around, but particularly in the lair itself, were numerous gnawed bones, so the lair had obviously been in use for a long time. On a nearby tree, at a height I could hardly reach with the end of my rifle, were several longitudinal claw marks. I showed Krylov where the bark had been stripped. I pictured the tiger emerging from its lair, yawning, looking around, and then standing up on its hind legs, stretching its front ones up the tree, arching its back, and sharpening its claws.[2]

There were many old tracks in the snow around the poplar, both large and small. It meant that the lair had first been used by one tiger, which was replaced by a younger one. What we could not tell was whether the tiger had left to go hunting, or had we scared it off? Most likely it went of its own accord, otherwise it would not have forgone the opportunity to feed on dogmeat, especially since the dog had conveniently delivered itself to the lair.

After examining the spoor Krylov wandered off. The longer I remained there alone, the more terrifying the situation felt. I could imagine the monster coming back. Seeing people present, it would stop and watch in a state of indecision, then would creep, keeping very low, along the dry streambed just beyond. Suddenly a sound made me start and turn around. On the next tree perched a crow.

Krylov soon returned, and we set off back the way we had come. Over the darkening forest spread a bright greenish-blue sky, tinted yellow and orange in the west. The forest seemed to fall even more silent, the snow took on a pinkish hue, and the air felt distinctly colder.

We soon reached the stream, but then took a wrong turning and arrived at the Anyuy half a kilometre higher up than our camp. We immediately felt relief. The forest - threatening and full of danger - was behind us. In front stretched the broad expanse of the snow-covered river. It was getting dark, and stars were appearing here and there. Night was bringing the exhausting day to a close.

In half an hour we were back at the camp. Everything was ship-shape. The tent was pitched, and from the stove pipe were issuing smoke and sparks. The draught dogs were resting on conifer branches and herbage spread for them. Supper was being prepared by the duty cook.

I spent the evening writing up my diary and other notes, then did whatever else was necessary before turning in early.

During the night I woke up. Opening my eyes I saw the stars through the openings in the tent. The silence all around was interrupted by a faint snoring. I also heard two dogs growling over something; but they evidently soon settled their difference, and the forest silence took over once more. In a few minutes I was asleep, but suddenly a loud noise woke me again. I lept to my feet, while my companions were also waking up and looking around, wondering what was afoot. From outside came yelps, barks and howls. The dogs were tugging at their tethers as if crazed. Marunich, Glegola and Vikhrov immediately put boots on and ran outside. I too dressed as fast as possible.

It was beginning to get light. Between large clouds shone a few remaining large stars. The land was still cloaked in darkness, although already the shapes of things could be made out - the snow-white surface of the river, mist over the ice-free stretches, and the trees, which in their winter apparel seemed to be dreaming, trapped in a frozen torpor.

My attention was immediately attracted to the dogs. Their straps were tangled, and they themselves were all in a heap and staring fearfully into the bushes. Some had broken free from their tethers and, with tails down, were running around the camp. What had happened? Why such a commotion?

'A tiger must have come,' said Krylov, emerging from the tent.

I ordered the dogs to be calmed and sorted out, a fire to be lit, and tea to be brewed.

'My dog's gone!' suddenly shrieked Marunich in a strange voice.

Krylov and I went over, and right there in the snow saw drops of blood and fresh tiger spoor. All was now clear. A tiger had attacked, and taken the dog. All the riflemen ran over to Marunich while I grabbed my rifle and made a short reconnaissance.

Remembering what Dersu had taught me, I carefully studied the spoor. It was not difficult to make out what had happened. The tiger had come along the bank on our side of the river. Seeing the camp, it stood for a while on a fallen tree partly embedded in a shingle bank at the bend of the river. It then lay down with its paws stretched out in front. After that it crossed over, using the fallen tree as a bridge, and crept up to the camp along a channel running parallel to the bank. This gully brought it close to where the dogs were tied up. It would have taken only a moment to jump over to the dogs, grab one, and get back into cover.

Eventually the dogs calmed down. Now it was the turn of the riflemen: they were very excited, all talking at once, and telling each other about the attack, who was asleep, who was awake and heard something, and who was dreaming what. Everyone had a different plan to find the tiger and recover the dog. Discussing the situation calmly while drinking tea, we decided to take four rifles and to track the tiger. If it had taken the dog to its lair, we would find it there; if, on the other hand, it had discarded its prey and run off, we would set up a rifle trap by the poplar, then come back to camp and await results. This in fact is what we did. At 6 o'clock in the morning Krylov, Rozhkov, Glegola and I set off, taking with us all the materials needed for setting up the traps.

The weather was calm. Mist over the open water was becoming thicker. Golden-pink rays of the rising sun flushed high clouds and the eastern slopes of the mountains with their frost-covered pines. It promised to be a magnificent day.

The tiger track led us straight into the forest. From the spoor we could see that on seizing the dog the tiger made a series of jumps, then trotted, and finally covered the next five hundred metres at a walking pace. The snapped off lead dragged in the snow. After a kilometre the track stopped in a small clearing, where the tiger had begun to eat the dog. Hearing our approach it left its prey and ran off. When we reached the dog, about half of it had been consumed. We began to set up the traps on the spot.

This is how it is done. Two stakes are inserted in the ground a short distance apart; alternatively two trees close together can be used, and this is in fact preferable since freshly planted stakes and the chips produced when cutting them could raise the suspicions of the animal. The rifle is tied to the stakes or trees, pointing at right angles to the animal's expected direction of movement. For a tiger the rifle is fixed at 17-35 centimetres above a man's knee height, depending on the size of the animal. Then the rifle is loaded and cocked. Between the trigger and its guard is inserted a wooden lever, to the long end of which is tied strong blackened cord, which is then extended across the track and tied to something on the other side - such as a bush, tree or rock. It is better not to stretch it tight since crows, of which there are always plenty around a kill, might perch on the string and trigger the trap by their weight. Tigers used to forcing their way through undergrowth pay little attention to the string: they just press on, stretch the string, and so get shot.

We set up four rifles so that, from whatever direction the tiger approached, it would pass one of them. When we had finished we used spruce branches to brush the snow and so remove our tracks before returning to camp. At dusk I instructed Vikhrov to get a fire going near the dogs, and post an armed guard. At night all was quiet: no shots were heard. Early in the morning Rozhkov and Krylov skied to the traps and reported that the tiger had come to its prey during the night. Giving the carcase a wide berth, it approached from the opposite side, but being suspicious it decided not to touch it. Stopping just two paces from one of the strings it lay down. Judging by the fact that the snow under it melted, it must have stayed there a long time. Before dawn it went away. Rozhkov and Krylov followed it as far as the lair, from which it ran off after hearing them approaching.

So we had to set up the traps around the lair instead. Krylov and two riflemen undertook to do this, and I stayed in camp. They returned as it was getting dark and reported that on the way back, on the bank of a stream, they found a pile of frozen fish, fenced in on all sides by interlaced willow sticks. The fish must have belonged to the Udege, Makha Kyalondiga. The tiger had broken into the wickerwork and enjoyed a fish supper. Judging by the spoor this must have happened the previous night.

When the light began to fail we collected all the dogs into the tent. Of course it was very inconvenient, but we decided to put up with it.

The next night was also quiet. The tiger skirted around the lair and also around the dead dog. It sensed danger and did not want to risk its life, but I decided to stay a week if need be - until hunger made it less wary. However, the tiger was cleverer than I thought. All day long we sat in the tent, each using the unexpected free time in his own way: the riflemen mended their clothes, the dogs' harnesses and the sledges.

In the late afternoon I dressed and left the tent. The day was drawing to a close. In the west the sky was pink, and the snow had a bluish hue. Ravines in the mountains were tinted soft violet, while small clouds on the horizon glowed red as if made of an alloy more valuable than gold or silver. All around was quiet. Mist was forming over unfrozen stretches of the river. Very soon large stars were lighting up in the sky, and night would soon be ruling the world. Just then I saw Makha running towards us along the ice on the river. Something had scared him.

'What has happened?' asked Kosyakov.

'*Gypy*,' he replied, '*Kuty-Mafa ono inaki dzyahzyangii* (Bad; tiger took a dog)'. He added that it had not gone far, and was eating the dog near his hut.

Only an hour and a half remained before the light would fail altogether. I calculated that if I ran for rifles, there and back would be two kilometres, then one and half kilometres to Makha's yurt,[3] and a bit more to the tiger, perhaps half a kilometre - the whole distance would require at least an hour. By then the tiger would have had time to eat the whole dog and walk away unhindered. Thinking a bit more, I asked Makha if he had a rifle. He said 'yes', but he himself would not shoot the sacred animal. On the other hand, since Kuty-Mafa had attacked his home, he promised to guide me to the tiger. I decided to accompany him, while Krylov and Rozhkov ran into the forest to fetch the rifles.

In only twenty minutes we reached the yurt. His wife was sitting with their two children in a corner, very frightened. She had lit a big fire at the entrance. Their dogs had been brought inside, and were tied to the poles supporting the bark on the roof.

Makha took his rifle out from under his bed, and handed it to me. It was an ancient Berdan rifle of cavalry type. I examined it and everything seemed in good order. I unlocked the safety mechanism and squeezed the trigger. The spring seemed to be weak, with insufficient force to operate the firing pin. I asked him whether it misfired, and how well it fired. He replied that it fired well.

Just then I heard behind me a whisper and caught the scent of the bog plant *Ledum*. Turning around I saw that Makha's wife had pulled out from somewhere a wooden human figure, about twenty centimetres high, armless but with bent legs and an elongated face with blue beads for its eyes. She suspended the sooty idol from a string over the hearth, addressed it as Kasalyanku, whispered something, and burnt *Ledum* leaves in front of it. I had no time to ask about the meaning of the idol, to which she entrusted the protection of the home and her husband against the savage beast.

I began pressing Makha to get a move on. He changed his footwear and fetched a spear. Before we set off, he was emphatic that it was not he who was going to kill the tiger; the spear was just in case....

I understood him and confirmed that all responsibility for the tiger's death would be mine and mine alone, and that his family had no part in it. Then we put on our skis and set off, with Makha leading. After about two hundred paces along the river we turned up a small side stream.

While Makha had been running to our camp, and when later I was examining his weapon and ammunition, something happened which we could not have foreseen. The tiger came closer to the yurt and, taking up position on the ice of the tributary under a steep bank, got to work on the dog.

Having no idea the tiger was so close, we proceeded along the stream, myself close behind Makha, talking freely to each other. Suddenly he stopped, and since I wasn't expecting it I went into him and almost fell over. I glanced at him. His face betrayed intense fear. Then I quickly looked in the direction he was staring, and saw the tiger about three hundred paces from us. It was standing stock still, like a statue, its front legs leaning against a tree trunk frozen into the ice, and it was gazing straight at us.

I thought it made a brief grimace, and for an instant bared its teeth. The fur on its back stood on end, but immediately settled again. Its whiskers seemed to be quivering. Twice the end of its tail slowly waved back and forth. It is in situations like this that a few minor details become stamped on the memory permanently and particularly clearly. I can't say that just then I was especially scared. More probably, like Makha, I simply froze to the spot and became totally absorbed in the scene.

Suddenly I remembered the rifle. The thought immediately occurred to me, 'I must fire straight away, or it'll be too late.' I pressed the stock to my shoulder and squeezed the trigger. There was a click but the rifle did not fire. Hastily I tried again. Meanwhile, about twenty-five paces behind me appeared Rozhkov and Krylov. Seeing me aiming at the tiger and not firing, they decided to open fire from a distance. But now something else happened that no-one was expecting: their rifles, after two days out in the frost and no doubt well greased, also misfired.

The tiger stayed put for about another two minutes, then turned and, occasionally looking back, slowly walked away into the forest. Just then two bullets whistled past us, but both missed the tiger. It made a great leap and in an instant appeared at the top of the steep bank. There it paused, once more gazed at its enemies, and disappeared into bushes.

All four of us went over to the spot where the tiger had, just moments before, been standing. There, on the ice behind a log, lay the half-eaten dog.

In some agitation I sat down. I had no idea what to do. I was still holding Makha's rifle, and in a fit of temper threw it aside, like a stick, into the bushes. When Makha realised that all three rifles had failed to go off, he became indescribably agitated.

I began cursing his Berdan, but he had a different interpretation of what had happened. In his eyes, the tiger had become even more sacred. It was omnipotent: it had only to look at a rifle for it to stop working. Knowing this, it just calmly watched its two-legged enemies approaching. How could one hunt such an animal? Makha was so convinced of this that without a word he picked up his rifle, blew the snow off it, and calmly returned along the ski track to his yurt.

We stayed, sitting on a log, and began discussing what to do next. It was decided to use the two rifles to set up traps beside the dog's carcase. Access from the third side was blocked by the steep bank, and the fourth - for which we had no weapon - remained open. Suddenly I had an idea: Makha had a cross-bow, so why not cover the fourth side with that?

Having set up the two rifles we brushed over our tracks and went back to the yurt along the river ice. Makha produced his cross-bow[4] and showed us how to set it. He repeated that he would take no part in the attempt on the tiger's life. Again I had to reassure him, saying that it was our affair entirely and he would not be responsible.

Krylov and I immediately went back to the dog. Covering the approach from the stream with the cross-bow trap, we crossed to the opposite bank and began to keep watch for the tiger.

It was getting late. The last traces of light illuminated the western sky. Bright stars came out one by one. It felt as though there fell on the Earth not only their cold pure light but also a somehow mysterious and indescribable sadness or melancholy. The darkness and silence merged with it, and in inaudible waves filled the ravines in the mountains, the quiet forest and dark air.

I sat behind a bush and dared not stir. Sometimes I thought I could see some kind of shadow on the stream. I was anxious and lost my composure - not from fear or from cold, but from nervous tension. At last it became pitch dark, and it was impossible to see the spot where the dog was lying. I raised my rifle and tried taking aim, but could not see even the sights.

'Let's go,' I said quietly to Krylov. 'It's freezing hard. I'm shivering.'

'My legs are shivering too,' answered the Cossack as he got up.

In half an hour we were back in the yurt. Everyone was asleep except for Makha, who was sitting by the fire waiting for us. Krylov and I heated water, ate some frozen fish, drank tea, then lay down on a bear skin and immediately fell fast asleep.

Makha's wife woke me at about six in the morning. She cooked us two pieces of fish with roe and *chumiza*.[5] Everyone else was up, and only I had overslept. My first question was, had anyone heard a shot during the night? Makha said he was awake all night, and heard nothing. I asked him to visit the traps, and he began to get ready. Meanwhile we had our breakfast.

In twenty minutes he returned in some agitation, and reported that the tiger had set off the crossbow. He hadn't stopped long by the dog and had not - as he should have done - examined the spoor. He simply noticed that the trip wire had moved and that there was blood on the snow. The news immediately raised our spirits. If seriously wounded the tiger would be lying down somewhere nearby. Hurriedly finishing our meal, Krylov and I dashed off with Makha to the spot.

It was a calm frosty morning, and the sun was just appearing over the horizon. The eastern slopes of the mountains were flushed pink, while the snow in gullies between the mountains still kept its twilight violet hue. By the opposite bank swirled mist as it formed in the frozen air over ice-free patches of water; it settled on the ice, bushes and stones as tiny iridescent drops.

When we arrived at the spot we immediately noticed the churned up snow and numerous, bright red patches of blood. Examination of the spoor revealed that the tiger had approached the dog not long before dawn. Avoiding the rifles it stopped, then, not seeing the other string, pressed on towards the carcase, so pulling the cross-bow trigger. The dart struck its paw. The tiger made a mighty leap, landed and, twisting around to its left, tried to pull the dart out with its teeth; but its paws were all the while slithering on the snow-dusted ice. As it writhed on the ice it fell off the edge into the water, but immediately pulled itself out. Here we found the shaft of the dart, which it had pulled out with its teeth, but the tip remained in the wound and was clearly giving the animal a lot of trouble. The spoor also told us that it spent a long time struggling to rid itself of the tip - lying on its side, twisting its body in a tight curve, and pressing its paws against any irregularity in the ice. Eventually it made its way over to the steep bank. Pressing its back against the rock, its front paws against a fallen tree, and its hind ones on the frozen shingle, it managed to grasp the tip in its teeth and wrench it out of its foot. This is when it lost most blood. Free of the dart at last, the tiger made off into the forest. To begin with it dragged its injured foot, but later trod on it gently.

The sun was now rising, and although there wasn't a cloud in the sky, the light was strange - whitish directly overhead but grey near the horizon. Our whole attention was concentrated on the spoor. The tiger walked on, choosing places with thick undergrowth and less snow. Its right paw left deep impressions in the snow, while the left one made only slight indentations. It was obvious that the tiger was being careful with the wounded paw, and putting as little pressure on it as possible. There was less and less blood, and finally we saw the last of it. This meant that the wound was not serious, and so we could not hope to catch it up soon - especially since the snow in the forest was not deep, so that the animal could move without difficulty.

As we went on the forest became denser, with more fallen trees and other debris. Venerable giants, unmoving and seemingly petrified, soared above the rest of the forest, at first as isolated individuals but later in whole battalions. It was almost as if they were closing in to help the lord of the forest escape from those upstart humans. Darkness reigned under the trees: even the bright light of the sky could not penetrate down below. The eternal deathly stillness was broken only occasionally by the wind, high up in the canopy. The sound felt like a warning threat.

At one point we were about to close in on the tiger. It had crept under a fallen tree and was lying on its side, evidently licking its wound. It was so preoccupied with this that it did not notice until we were almost upon it. In such situations one has to be extremely cautious.

'Captain,' Makha whispered to me, 'no need to hurry.'

Just then I heard a noise and spotted the large striped body as it slipped away. Rozhkov also saw it, but neither of us managed to fire.

It was now about midday. We had a rest before going on, and chewed dry salmon meat, which Makha had thoughtfully brought with him. Discussing the situation we decided to continue the pursuit, although Makha was not enthusiastic. He said that it was no use continuing, since the wound was only slight, it was on the mend, and far from avoiding the forest debris the tiger would make for even more impenetrable places.

His reasoning was persuasive. Even so we resolved to try our luck. After our scanty meal, and swallowing snow to quench our thirst, we set off once more in pursuit of the tiger. From its track we could see that it made a series of jumps, then resorted to trotting before settling down into a walk.

Now we left the floodplain forest, and entered old pine forest. The massive trunks soared above us, like columns in an old ruined cathedral supporting the vaulted ceiling high above. Down below between them were some younger trees. The branches of the old and young trees interlaced, and the trees were thickly entangled with lianes forming an impenetrable barrier. The primaeval forest and its dense silence made us feel trapped. A green half-light, like some kind of unbearable mystery, pressed on us all around, and I couldn't help thinking it would be better to go back before it was too late.

Krylov was ahead. Suddenly he stopped, pointed to the ground, and said quietly, 'The tiger has gone around behind us.'

It was true: running obliquely across the track we were following was another identical track. The tell-tale lameness of the left paw identified the animal. It was clear that, losing patience with its persistent pursuers, it had decided to go on the offensive. So it executed a large loop and, twice crossing its earlier track, decided to set an ambush. This was serious. Slowing our pace, we became extra vigilant. In another five hundred metres or so the tiger tracks crossed a third time.

We were so absorbed in the hunt that we failed to notice the failing light and the chill of the approaching evening. High above, through gaps in the canopy, could be seen dark clouds.

'*Manga*,' said the Udege, glancing up. '*Tamna. Imasa cagdy bi...* (Bad. Clouds. There will be much snow)'.

In the forest it became very dark. The tree trunks seemed to move closer together to block our way. It was as if the forest, knowing it could not, by itself, stop the men hunting down one of its own, was calling on the elements for assistance.

I glanced at my watch. It was about five o'clock. Soon the sun would drop below the horizon and dusk would fall. Weighing up all the pros and cons, we decided we had to go back, if only to satisfy our hunger, which was making itself increasingly felt. We were also feeling something else - bitterness and disappointment at our lack of success. Krylov and Rozhkov cursed the tiger and its tracks. It had to be admitted, circumstances were forcing us to 'cede the field' to our enemy just when it was about to fall into our hands. But the voice of reason was decisive: however valuable a tiger skin may be, ours were dearer to us.

'*Imasa bi* (Snow is coming),' observed Makha.

I looked up. The sky was rapidly getting darker, and seemed to be falling towards us. It was an unpleasant and eerie sensation. I felt a few snowflakes on my face.

Just then a gust blew through the canopy. The forest stirred, became noisy. The wind struck from another direction, and as suddenly died away. The air was filled with a whitish murk, blanketing out the surrounding hills as well as the ground and everything else that had been visible. The wind came in gusts, and between them could be heard the rustle of the snow falling in large flakes.

'We must hurry,' said Krylov, 'if we are not to spend the night in the forest.'

We quickened our pace, but that did not speed our progress. The ski track quickly filled with snow so that it could hardly be made out. Soon the track we had come by disappeared altogether. Krylov halted. Makha took the lead. He knew the area well, and chose a new direction across snowdrifts and in a straight line towards his yurt. The light failed, so that it was difficult to make out anything ten paces ahead. The snow and wind, as if in cahoots, threw themselves viciously against anything that got in their way. Now and then a spruce tree could be made out through the darkness, with its branches frenziedly swaying back and forth. Now would unexpectedly loom out of the darkness some stout trunk, dusted in snow, and now a fallen tree or a dead standing stump, quite branchless. As we pressed on, these apparitions passed us and disappeared without trace in the gloom.

The Udege soon brought us to an old river channel, which I thought was leading us in the wrong direction. As if reading my thoughts he turned to me and calmly said, 'Not far now.' In about fifteen minutes the channel led us to a path, which turned sharp left. Here the gusts of wind became stronger. I began to shiver, and became concerned about my riflemen, who had dressed casually to make it easier to track the tiger on their skis.

It became ever harder to make headway. We were hindered by a head wind and loose snow, into which the skis sank deeply. I had completely forgotten about the tiger, and thought only about reaching the yurt as soon as possible. Even by straining every nerve I could hardly drag my legs along. My arms were shivering violently and my face was frozen. Just when things seemed to be getting impossible we came to the spot where the rifle traps had been set up around the dog's carcase. It raised all our spirits. Just ten more minutes of ski-ing, and through the trees I could see reddish smoke and sparks rising from the yurt roof.

Thank God, we were safe! Having learnt from bitter experience, I took a handful of snow and, with my back to the wind, began rubbing the snow on my face. It was unbearably painful, but I rubbed hard and long in the knowledge that it was absolutely essential.[6] The numbness of my cheeks began to diminish, and my fingers also became more mobile and recovered their feeling. Only then was it sensible to enter the warm dwelling.

A fire was burning fiercely inside, radiating heat and light. Makha's wife immediately began cooking supper, while we brewed tea and relished rusks. After supper Makha spread out bear skins for us. We took our boots off and lay down. It seemed the height of luxury. I drank tea and listened to the blizzard raging outside.

'Thank goodness we turned back in time,' I said, expressing my thoughts out loud. 'It would have been quite impossible to continue stalking the tiger. The blizzard would have obliterated the spoor, and we'd have lost precious time and suffered from the weather to no purpose.'

Makha interpreted the situation in his own way: the tiger is a sacred animal; the forest and mountains protect him; we must never try to fight him with guns; the only way is to use shaman ritual, and ask the tiger to go away.

Krylov tried to object, but Makha would not budge. He cited a whole series of facts to prove his point: three rifles misfired, the tiger's dart wound soon healed, the forest tried to give it cover, creating all kinds of obstacle for us, and when we had almost succeeded, suddenly a blizzard blew up, the wind wiped out its spoor and we were forced to retreat. The Udege was firmly convinced that if we had continued pursuing the taboo animal, there would have been some catastrophe, such as our getting lost in the taiga and dying from hunger and exposure.

Listening to him, I began to doze.

Strong gusts shook the yurt to its foundations. The blizzard howled viciously through the forest like a herd of demented animals roaring along to some unknown destination. Near the yurt some tree or other was swaying and squeaking. On the roof a piece of bark was making a variety of noises according to the strength of the wind. Lulled by these sounds, still chilled to the bone and worn out by the hard skiing, I fell into a deep sleep.

Sounds woke me during the night. Opening my eyes I saw that the fire in the yurt had died down: only a few smouldering embers were left. Directly opposite me Makha was sitting on a birch mat, holding a shaman's tambourine and rattle. He was singing something doleful and at the same time pressing the tambourine to his mouth, making his voice, resonating from the taut skin, now swell and now fade to a whisper. While singing he gently tapped the rattle on the tambourine. In front of him, hanging on its string, was that same wooden idol with the eyes of blue beads, before which his wife had burned *Ledum* leaves the day before. Gradually his song turned into a recitative. Then he got to his feet, took his axe, two small wooden pegs and an iron pan, into which he placed four embers from the fire, and went outside. I rose and followed him.

Outside all hell was let loose. It was pitch black. The wind almost knocked me off my feet, and it felt as though somebody were throwing snowballs at my face. The forest was howling, and the noise conveyed feelings of displeasure, complaint and threat all at the same time. I soon became accustomed to the darkness and looked around.

The Udege threw the two pegs into the snow by the entrance to the yurt, then walked around it. At each corner he turned towards the forest, shouted 'e-ye' and threw an ember. After that he went back inside, took down the blue-eyed idol, shoved the tambourine behind a piece of bark, and threw fresh wood onto the fire. Returning to his former seat he wiped his face with his hands and started to smoke. I realised that the shaman ritual was over, and began brewing tea.

About five minutes passed in silence. Then I began talking about the blizzard, but switched to the tiger, and then cautiously broached the subject of the wooden idol. I asked him the meaning of the ritual and the embers.

'Oh, the idol, it is symbol of the spirits,' he replied.

Then he told me how he felt guilty before the tiger on three counts. First, he had told me about the dog taken by the tiger. Second, he gave me his rifle and crossbow. Third, he went with me to pursue the wounded tiger. He was asking Kasalyanku to protect his home from the tiger, and so intoned incantations over the pegs, which would bar the way to the yurt. He surrounded the yurt with sacred fire on all sides. Now he would be able to hunt as before, quite safely. Finishing his smoke, Makha got ready to sleep. I did the same, but lay awake for a long time.

After about two hours the storm abated somewhat. The wind died down, and the gusts became less frequent. Through the smoke hole in the roof could be seen the dark cloud-covered sky. It was still snowing, but other forces were gaining the upper hand, and were about to quell the raging elements and so restore the world to its normal order.

Chapter 7

KAPLANOV

'To anyone unaware of Lev Georgiyevich's rich experience earlier in life, it might seem surprising how easily, simply and, one might say, without hesitation Nature would open page after page of its book to him, as to few others.' [1]

Among Russian tiger biologists one name stands out above all others - Kaplanov.[2] This was a man who showed great promise in his early years, who fulfilled that promise when he was given the chance, and yet whose life was tragically cut short at the age of thirty-two.

He was born in Moscow on 7th November 1910, while his parents were still science students - his father in physics and his mother in zoology. Although christened Lev, in the family and among his friends he was known as Leva. It is a curious coincidence that in Russian '*lev*' means 'lion'.[3] His father, Gazaros, was an active man, with an encyclopaedic knowledge in many fields, and he encouraged Leva to take an interest in science. His mother, Sofya, had numerous teaching and other commitments which often took her away from the family; even so she also influenced Leva, particularly in a love of nature.

The family led a rather simple way of life, with few luxuries. Leva was an obliging and unpretentious boy, spending much time reading and also studying languages. Formal education did not suit him, so he largely taught himself while bringing home bad reports from school. It came as no surprise that his formal education ended with secondary school, though it was puzzling why such a talented youth should not want to enter higher education.

Even as a child Leva was unusually observant, and was quick to learn the names of animals and plants. He loved excursions into the forest, at first with his mother and younger brother Andrik, and later alone or with his close friend Vadim Rayevski,[4] who was generally known as Dima. When Leva was twelve or thirteen years old, he and Dima were fascinated by small mammals - mainly rodents, shrews and bats. From the age of fourteen, Leva was in the habit of recording his observations regularly, and made collections of mammal skins and bones. It was already clear where his main interests lay.

During these years he received encouragement not only from his parents but also from Professor S.I. Ognëv and other biologists in his laboratory at Moscow State University. Other help came from the staff at a museum, and from the Young Biologists' Club at Moscow Zoo. From the initials of its Russian title, the club was known as *KYuBZ,*[5] and its members were *kyubzovtsy*. As well as observing and helping to look after the zoo

Lev Kaplanov
1910-1943

animals, the *kyubzovtsy* were taken on excursions into the surrounding countryside. Later Leva came to know the noted animal ecologist Professor Alexandr Formozov,[6] to whom he continued writing letters later from the Russian Far East. The young Kaplanov was unusually lucky in having so much expert guidance and help in his desire to become a zoologist, despite dispensing with a university education.

Nothing dramatic occurred in his life until spring 1925, when he and Dima were invited to join a party from the museum to the forest of Tver Region, which is to the west of Moscow. This was Leva's first real expedition, the main aim of which was to hunt bears. Gazaros had been anxious about letting the boy go, considering him too young to take part in a serious hunt, but in the end he relented. Perhaps he had a premonition, for only a few days later a telegram arrived: 'Inform hospital, eye injury'.

Apparently Leva had tried to remove a live cartridge from a gun, and it had exploded in his face. Arriving in Moscow, he was taken straight to the hospital. When the bandage was removed, it was clear that one eye was a complete mess, and there was no hope of restoring its sight. An operation was carried out, but it was not known for another two months whether the sight in the other one could be saved. Leva was surprisingly unperturbed by the accident, and by what the future held in store for him. Every day he and Dima would look after their captive shrews - including water shrews - and went around their traps in the hope of catching more. Only when the autumn came could the family be sure that the danger was past, and Leva would not go totally blind.

During his teens he lived life to the full, as if the accident had not happened. In particular he fell in love with the taiga, and trained himself to go on long treks, studying the mammals of various areas around Moscow. He would stay out in the wilds for days at a time, practising endurance and what we would now call survival skills. Professor Formozov accompanied him on a trip to Kostroma Region to the north of Moscow in the autumn of 1931, and the local hunters were amazed at the young Kaplanov's willingness and ability to stay out during long October nights in bad weather on bogs and areas of burnt forest, to hear the roar of rutting elks.

After leaving school he found jobs in which he could work with animals. In 1931 he was an assistant at the Institute of Hunting & Fur Animals, working under Formozov, who persuaded him to enrol as a student in the institute. However, Kaplanov did not attend the lectures regularly, and so was soon excluded from them.

Siberian taiga (1933-36)

Apart from a few months spent in the Caucasus, Kaplanov's early experience of wild country was acquired in the Moscow area. His chance to become acquainted with the Siberian taiga beyond the Urals came in 1932, when he was able to spend the summer studying an area with a view to introducing muskrats. At this time there was a serious drive by the Soviet government to 'enrich' nature - in particular by enhancing natural resources, such as populations of fur-bearing animals for trapping. There were numerous acclimatisation and introduction schemes involving muskrats, beavers, desmans, musk deer, raccoon dogs, squirrels, and even sea otters.[7]

This first visit to the Siberian taiga was inspirational. It was no surprise, then, that Kaplanov seized the opportunity to work as a biologist at a research station on the River Demyanka - a tributary of the River Vasyugan, itself flowing into the great River Ob. In September 1929 ninety muskrats had been released, and over the next three years another biologist had been studying their progress. He reported rapid reproduction and estimated the population at about 10,000 in 1933. This apparent success led to the establishment of a commercial trapping enterprise, but it soon failed since it turned out that the estimates of population size were wide of the mark.

There can be no doubt that the animals had bred successfully for a time, soon after the introduction. But then numbers declined, partly on account of disease. Kaplanov investigated the disease and other causes of mortality among the muskrats in 1934. Another reason for their lack of success was that many were being killed and eaten by pike and birds of prey.

It is a feature of Kaplanov's career that, with few exceptions, the mammals at the centre of his interest gradually climbed the scale of size - from the small rodents and shrews of his boyhood to the much larger muskrats. Now, with that work completed, he took another step up the scale. His focus shifted to the largest mammals in the district - elk and reindeer.

Apart from tracking the elk, he also tamed a couple in order to try domesticating them.[8] One he called Mashka, which had been fed by hand from an early age and was quite tame. Mashka would follow Kaplanov faithfully, wherever he went in the taiga. When he spent the night in the open Mashka would browse nearby, never letting Kaplanov out of her sight. To avoid her being shot by mistake by other hunters, he tied a bell around her neck.

Perhaps the greatest gains from Kaplanov's three years on the Demyanka were confirmation of his love of the taiga, and the development of his supreme skill as a *taëzhnik*, or person 'at home' in the taiga. In Formozov's words, 'By this stage in his life, through persistent effort he had already acquired those qualities essential for spending weeks or months successfully in the taiga, painstakingly collecting valuable facts about the lives of the most wary and secretive animals. Here on the Demyanka he convinced himself that, despite the loss of one eye, he could easily endure all the privations that come with a nomadic life, that paths were open to him as to few others, and that he could extract the maximum information from the hidden corners of nature'.[9] These extended trips in the taiga, often alone, involved carrying heavy loads, and so required exceptional fitness. But he loved it! As he wrote on the cover of his diary for 1935-36, 'These have been the best days of my life - my "golden" days'. And in a letter to Formozov, 'The world around me is a beautiful symphony of shapes and colours; far from life in the forest being boring, with each year, month and day the joy of living grows stronger and the more magnificent become the rivers, lakes and bogs, the dark riverside conifers, the lilac-coloured birch groves and the pink-tinted flooded areas. Everything moves me, and brings moral satisfaction and peace of mind'.[10]

Sikhote-Alin Mountains (1936-41)

In March 1936 Kaplanov was back in Moscow, with his dog Serko. Later in the year he received an invitation from K.G. Abramov, the director of the zapovednik that had been founded the previous year in the Sikhote-Alin Mountains: would Kaplanov like to join the staff as a zoologist?

So began five years spent working in an even wilder area than the West Siberian taiga. As a biologist his main tasks were to investigate the ecology of three large mammals - elk, *izyubr* and tiger. But before that was possible the unexplored areas needed to be explored, and a minimal infrastructure laid down - such as paths, huts and food stores.

This was known as the 'heroic age' of the zapovedniks. Some were literally unknown territory and so large that it could take weeks to reach the interior, or to trek out to the nearest settlement for supplies and trek back again to continue observing the wildlife or surveying the vegetation.

Even for somebody with Kaplanov's skills and fitness it could be a hazardous business, and occasionally he too was caught out. A case in point occurred in February 1938. He had set off, without skis and with as light a load as possible on his back,[11] to find the elks' wintering grounds. He became stranded by unexpected heavy snowfall. For several days he dragged himself along through deep loose snow, progressively exhausted by hunger and fatigue. On the way he discarded most of his load. Eventually, practically crawling on all fours, he came to a house. He was almost at death's door, and could do nothing for several days. Eventually, when fit enough, he made himself a pair of skis, and so managed to return to base only after twenty days.

Kaplanov first came across tiger spoor in the Sikhote-Alin in February 1938. He immediately felt an urge to apply his skills to the challenge of studying such a rare and interesting animal.

During the next winter he was still studying elk, and his dog Serko was always with him. It happened that some tigers were hunting the same elk that Kaplanov was observing at one site, and a large male tiger passed close by his hut several times. Knowing that Serko would go after a tiger just as boldly as it went after elk, lynx and bear, Kaplanov took the precaution of keeping Serko in the hut at night, while he himself always took a rifle when away in the taiga.

Then, one day in December 1938, Kaplanov went out early in the morning to fetch water from the river. Serko was with him, but suddenly bounded away and disappeared into a thicket. Soon there came a short yelp, and all fell quiet. Kaplanov ran to the hut for his rifle, and followed Serko's tracks. The body was soon found. The tiger had bitten into the dog's head, and its fresh tracks went off along the frozen river.

The sudden death of this wonderful dog, companion on many successful hunts on the River Demyanka, around Moscow, and in the Sikhote-Alin, upset Kaplanov deeply. He tracked the tiger a long way, but without catching it up. Then, knowing that sooner or later it would use its track along the river again, Kaplanov set up a rifle trap on the path[12] - positioning a rifle with a tripwire to the trigger - in the hope that the tiger would shoot itself. This is not an easy trap to set at the best of times, and Kaplanov did not even have strong twine of the necessary kind. However, set it up somehow he did, and later a shot was heard. The tiger had indeed triggered the rifle, and shot itself in the hind quarters.

Not every hunter would pursue a wounded tiger in the taiga alone. Kaplanov however tracked it for two days. Several times he almost had the chance to kill it, but each time the wary animal got away. While continuing the pursuit by moonlight he glimpsed the tiger very close by, apparently annoyed and lying in wait to spring an ambush. Realising at last what a dangerous situation he was in, Kaplanov wisely abandoned the hunt. The tiger's wound healed, but it was always lame after that, and so its tracks could easily be recognised when Kaplanov saw them during the following year.

Tracking tigers

Years of life in the taiga, with all its privations, took its toll even on someone as robust as Kaplanov. There were times in his early years at Sikhote-Alin Zapovednik when he sank into depression and had no energy. During prolonged treks on his own he was prey to loneliness, and there were bound to be consequences of the occasional serious physical and nervous exhaustion - as in the incident of February 1938.

In the previous thirty to forty years there had been no original studies of the Amur tiger, and with each year its numbers were decreasing. Kaplanov knew that there was no time to lose, and so with the agreement of the zapovednik director he began his study of the tiger in 1939, completing it in 1941. This was his most challenging project to date, but he was also probably the best qualified biologist to undertake it.

Fortunately, by late 1939, he was back to his usual self, as he noted in a letter of 25th January 1940: 'The years of depression and melancholy are past, and once more I'm full of energy and enthusiasm. All my earlier work seems merely to have been preparation for the present task. The course of my life has been a direct path to the stage I am now at. While working here I have been patiently and methodically preparing to realise my long-cherished dream - to follow the tiger's tracks and to become acquainted at first hand with the animals themselves'.[13]

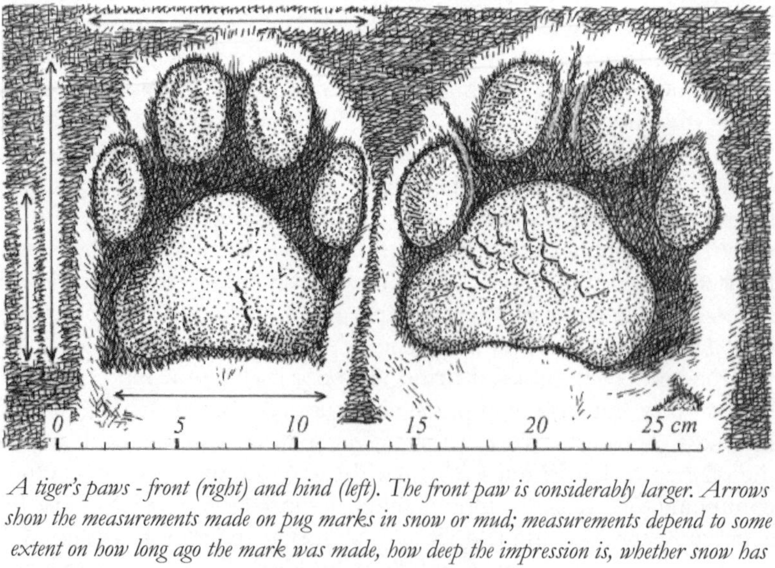

A tiger's paws - front (right) and hind (left). The front paw is considerably larger. Arrows show the measurements made on pug marks in snow or mud; measurements depend to some extent on how long ago the mark was made, how deep the impression is, whether snow has since melted, &c.

Part of his work involved collecting information from local residents - in particular hunters and teams specialising in capturing cubs for zoos and circuses. The other main approach was his own field work, often carried out in the company of another member of the zapovednik staff. His favourite companion was the guard Fëdor Kozin, son of the famous tiger hunter and cub catcher Aleksey Kozin. Field work was much easier than when Kaplanov had first arrived, because now there were adequate huts, food stores at strategic points, and a network of paths. So they were never too far from shelter, and did not need to carry heavy loads every day.

The great advantage the Russians had in those days over other tiger biologists was that so much can be learned from spoor and other signs of the animals in winter. With snow on the ground, but without more falling, there is really no way in which terrestrial animals can conceal their movements, and skilled trackers like Kaplanov made full use of this record in the snow. It also meant that much more could be learned about tigers in winter than when the snow had gone.

Spoor, when the impression is clear, gives a surprising amount of information (see diagram). The fullest set of measurements on a pug mark would be, in centimetres, the width and length of the whole footprint, and the width and length of the 'heel', in that order. Sometimes the fourth measurement was not made. Together with the length of the pace, depth of the impression, and other signs such as blood spots, these measurements might be enough to tell one tiger from another, and also its sex, approximate weight and age, bodily condition, and direction, gait and speed of movement. If it lay down in the snow, one would find out the length of its body.

Other signs - some of which can be used in summer - include scratch marks on tree trunks or in the soil, remains of prey, faeces and urine marking, and twigs chewed by cubs. So, if a tiger caught and killed prey, or had a fight with a bear, the exact sequence of events could with luck be read from the signs. It takes years of experience and uncommon skill to extract the maximum information from spoor and other signs, but Kaplanov was one such tracker. It was due to his tracking skills, fitness and dedication that he learned so much about the Amur tiger in his brief two-year study.

Kaplanov gave nicknames to 'his' tigers. For most of January 1940 he was tracking two animals, a male and a female. From the spoor size and length of pace it was easy to tell them apart. He called the male Zhestoki (Cruel) and the female Malenkaya (Small). From Malenkaya's elegant little footprints, Kaplanov thought that it was a small tiger, but later found where it had lain down in the snow, and so discovered that it was in fact quite large, and probably old.

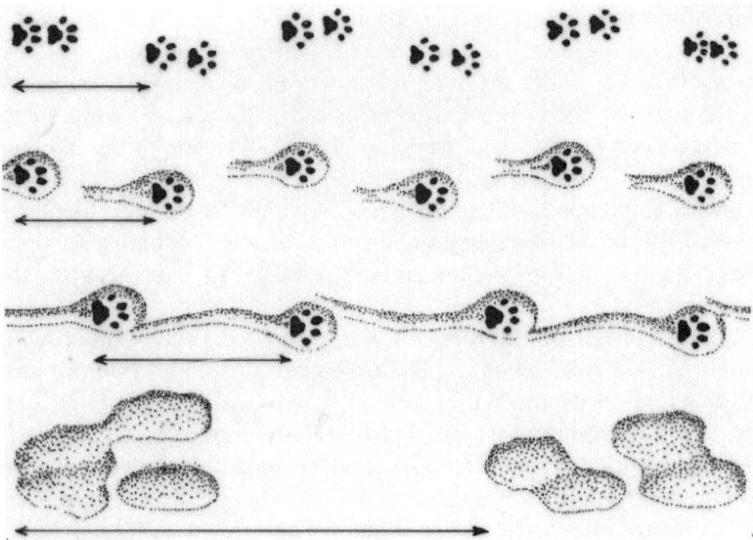

Spoor.
In the top row the tiger is walking slowly, lifting both left legs together, then both right legs. This unusual pacing gait is typical of camels and giraffes.[14] *In each pair of pug marks the left hind foot is placed in front of the left fore foot print, and similarly for the right legs. The top row of spoor is typical of mud or shallow snow, but in the second row, in snow more than 10 cm deep, the hind foot is placed on top of the print made by the fore foot on the same side (sometimes leading to five-toed pug marks). The walking pace (arrow) is 55-80 cm in adults. The third row of spoor is made by a trotting tiger, with a pace length of 80-110 cm, and much longer 'tails' (grooves in the snow made by the foot being swing forward). At the bottom are the rather indistinct impressions made by a tiger leaping forward when attacking prey: the hind feet land in front of the front ones, and the pace length can be from 2.5 to 5 metres. (After Yudakov & Nikolayev, 1987.)*

Another large female was Skrytnaya (Secretive). He also saw the tracks of the tiger he had lamed, and named it Kosolapy (Clumsy or Pigeon-toed) - a name often used for brown bears. A tiger with very large spoor - Moguchi (Mighty)[15] - seemed to have wandered into the area from elsewhere. About the time this spoor was first noticed, villagers on the River Iman reported seeing a massive tiger. One old hunter said that he saw on its head an unusual dark pattern. According to him such patterns, stretching from the eyes to the back of the head, used to be considered a sacred sign by hunters. It was said to mark an exceptionally strong animal bearing the name 'Veliki Vladyka' (Great Lord).

As an example of what he could read from the snow and other signs - or from the pages of the book of nature, as Formozov expressed it - Kaplanov described the life of a tigress and her three cubs from about

24th December 1940 to 15th January 1941.[16] He located the litter on 9th January on the floodplain of the River Kema, about 46 km from the sea. The cubs were several months old, each probably weighing 20-30 kg. It is likely that they had been born nearby, since two of the zapovednik guards had heard a tiger uttering three calls in rapid succession just five kilometres away, higher up the River Syao-Kunzha - a sound somewhat like an *izyubr* call, and sometimes made by a tigress when it is anxious about the presence of humans. Spoor had been noted in the area throughout the spring and autumn, and again on 11th December, just 30-60 m from former forestry huts that were now being used as a zapovednik guard post.

The right bank of the Kema just here carried mixed forest of Korean pine and deciduous trees, while on the slopes of the nearby hills was pure pine forest. At the foot of the slopes, at the edge of the floodplain, grew dense thickets of horsetail *Equisetum hiemale*, which attracted large numbers of wild pigs and *izyubr*. On the left bank, the hill slopes fell steeply to the river; the original forest had been burned, and was now replaced with mixed deciduous woodland of aspen, birch, oak and hazel.

For most of the time, the cubs remained on the right bank of the River Syao-Kunzha amongst pine, about a kilometre from the huts. The tigress was wary, and generally did not stray far from the cubs. She was in the habit of going up to the huts, apparently to check whether people were there - and there was no-one from 20th December until 9th January when Kaplanov arrived.

Needing food, the tigress left the cubs and went off hunting. She killed a three-year-old female wild pig several hundred metres away, and then returned for the cubs. She led them to the prey, and they followed in her footsteps. Her spoor measurements were 12, 13, 10 and 9 cm (as described above), while theirs were no more than 9.5, 10, 7.5 and 7, with a pace length just half hers. Her legs cut a furrow in the deep snow, and along the sides of the furrow could be seen where the cubs had brushed against the wall of snow.

Wild pig (or wild boar)
Found from western Europe all the way across to the Pacific, wild pigs form an important food source for tigers. Shown here are a boar (with tusks and prominent bristles over the shoulders), piglet, and sow.

The pig was lying in a thicket of bird cherry and other bushes, with plenty of fallen timber, so that the spot was well hidden. Around the pig was a well trampled area, and from it paths led to the stream - here not frozen, so that they could drink - and to their lair. The trampled area was the cubs' playground, and nearby was a tree trunk scored by the cubs' claws.

Just 200 m on the other side of the Syao-Kunzha, the tigress next killed a piglet, and brought it to the trampled area. She carried it in her teeth, and alongside her track could be seen marks in the snow made by the piglet's dangling legs. Practically everything was eaten, leaving only the legs, skin and some bones from the two animals.

Now the family moved home, the tigress taking the cubs across the Kema to the left bank. There they climbed the steep south-facing slope, and rested in the sun. The cubs played, fought, slid down steep slopes in the snow, raced back up and down again, and sometimes went off in different directions for a while. Later they all went down to the river

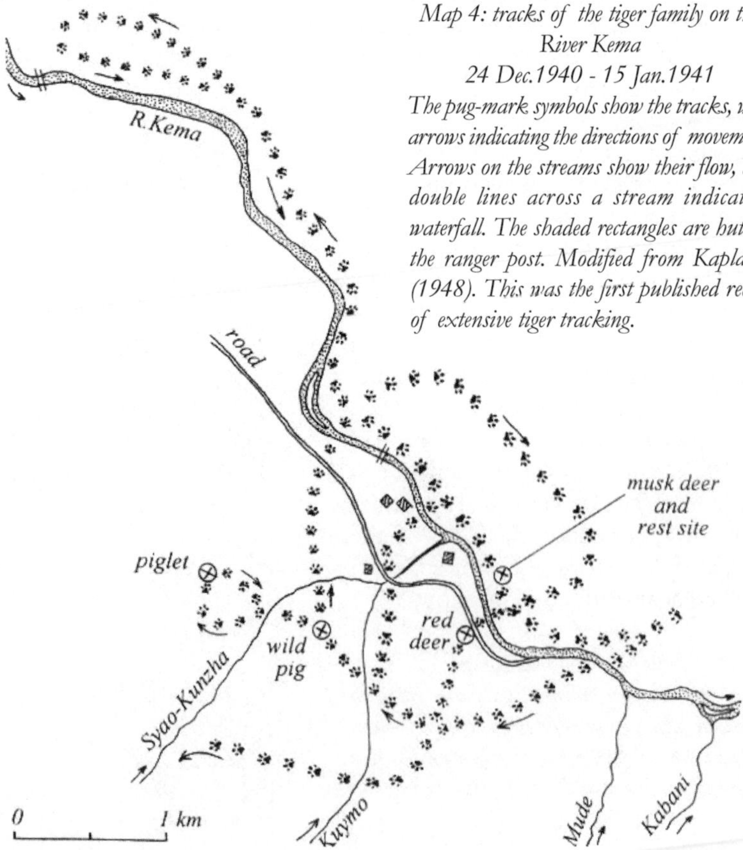

Map 4: tracks of the tiger family on the River Kema 24 Dec.1940 - 15 Jan.1941 The pug-mark symbols show the tracks, with arrows indicating the directions of movement. Arrows on the streams show their flow, and double lines across a stream indicate a waterfall. The shaded rectangles are huts at the ranger post. Modified from Kaplanov (1948). This was the first published record of extensive tiger tracking.

once more, but to a point further downstream, below the mouth of the Syao-Kunzha. There they rested in the bushes, and the mother probably suckled them. Nearby she caught a musk deer and the cubs consumed it, leaving nothing but a few shreds of skin.

Next the family wandered around on the Kema floodplain, and went upstream about three kilometres along the left bank to a waterfall. Here they could go no further, so turned back. Near the huts, in old pine forest on the right bank of the Kema, the tigress killed an *izyubr* stag on 6th or 7th January. The kill was made in a spot lacking cover, so she dragged it more than 10 m to some fallen trees. It could be seen where the tigers had lain down under the tree trunks. The area was well trampled, and about 50 m away was another trampled area, well hidden by undergrowth and fallen trees. Here many branches had evidently been chewed by the cubs. This was their playground, and where they were suckled.

When Kaplanov found the stag on 9th January, it was about half eaten. During the night the tigers returned and consumed most of the rest of it. A few metres away were three piles of faeces, probably deposited by the tigress.

Scared away from the stag, the cubs went off up the hill to the south. On 9th February the litter was noticed 12 km higher up the Syao-Kunzha.

Quite apart from the remarkable detail with which a tracker like Kaplanov could reconstruct the sequence of events, he also noticed something even more remarkable. Near the cubs, with the tigress in close attendance, were living herds of wild pig as well as several solitary pigs and *izyubr*. The tigress and cubs apparently did not frighten them, even though she made her kills only a few hundred metres from the cubs. Kaplanov saw pigs and piglets feeding quietly only half a kilometre from the cubs, and explained it by suggesting that the tigress hunted at night so skilfully and silently that she was not detected.[17]

From this and other case studies, Kaplanov was able to describe how tigers spend the winter. He also estimated that an average adult tiger kills and eats about thirty large animals each year, averaging about 100 kg each, in other words a total of about 3000 kg (three tonnes) live weight, or the equivalent in smaller animals.

The tigress and cubs, whose behaviour over twenty days has been described, consumed approximately 150 kg of *izyubr*, 100 kg of wild pig, 20 kg of piglet and 10 kg of musk deer. However, even taking this amount, the tigers living at the time in the area - and Kaplanov estimated the number as 12-16 adult tigers in the zapovednik, which then covered about 18,000 square kilometres - would have no significant impact on the populations of hoofed animals, assuming that their natural rate of increase was maintained.

The main preoccupation of Russian game biology during the 1930s was the management of game animals in the wild so as to maintain adequate numbers for hunting and trapping, and to a lesser extent also sport hunting. Kaplanov himself was not averse to shooting animals for research purposes, to sell the furs, or simply for trophies. As we have seen, he was prepared to set a trap with the intention of destroying the tiger that killed his dog in December 1938. Yet by March 1940 he was writing in a letter to Formozov that 'Although the honour of acquiring a striped pelt is very tempting, I think that preserving wild free tigers is far more honourable. This I am trying to do, and my ambition has already been satisfied by making the closest acquaintance with tigers. There is no obstacle to killing tigers in the zapovednik if I wanted to, but destroying just one of these beautiful animals would be contrary to the conviction that now consumes me'.[18]

He seems to be saying that, while he might well have yielded to the temptation just a year or two earlier, he felt now that it was wrong. He had managed to break away from the old hunting ethic, at least so far as tigers were concerned. In fact he was already writing, with the zeal of the converted, to Formozov in January 1940, how 'I consider it our sacred duty to protect the wild tiger in freedom in Sikhote-Alin Zapovednik for future generations, as one of the supreme wonders of nature. I hope that people in our communist society will have the benefit of seeing - on a par with the greatest achievements of technology - the tracks of the giant striped cats [...] in the snow of the Sikhote-Alin Mountains. Here I am trying to provide a solid scientific foundation for this task. And now [...] I appeal to you, and to all Moscow zoologists, to help me put a stop to the capture of tiger cubs in the zapovednik'. He went on to say that he would refuse to shoot a tiger for the All-Union Agricultural Exhibition, and had given up his dream of hunting it; '[...] "bloodless hunting" will be just as interesting, and more useful both to science and to the tigers themselves - the last tigers of the Sikhote-Alin'.[19]

Moscow (1941)

In the spring of 1941 Kaplanov was back in Moscow. Germany had attacked Russia, and so he took his turn to help with anti-aircraft protection on the roof of a block of flats. He was not called up on account of being blind in one eye.

Three significant events occurred during his brief stay in his home town. First, his years of loneliness on long taiga treks were now over, because he fell in love with and proposed to a young woman called Lida Kastalskaya, and was accepted.

The second event was his chance to publicise his message about conserving tigers. He gave a presentation on his work and ideas to a meeting of the Moscow Nature-protection Society at the University Zoology Museum. Although it was well received by the audience of zoologists, it also provoked some adverse reaction in the press: nature conservation in the modern sense was not automatically considered in those days 'a good thing' among all sectors of society, and certainly not among traditional hunters.

The third event was initially a disappointment. Kaplanov had set his heart on returning to Sikhote-Alin Zapovednik to study sable, but the Main Zapovedniks Administration had other ideas: they posted him to the neighbouring zapovednik, then called Sudzukhe, and more recently named Lazovski Zapovednik.

Sudzukhe (1941-43)

On 12th September 1941 Kaplanov and Lida left Moscow to start the long journey back to the Russian Far East. Soon the zapovednik director was called up to the front, and Kaplanov found himself taking over.

His main task as a scientist was to study the curious antelope-like mammal, the goral.[20] It was being much persecuted by poachers because the meat, and particularly the goral embryos, were considered to have medicinal qualities, and so fetched a high price. Accordingly much of Kaplanov's time was taken up with organising and enforcing protection,[21] and trying to maintain the zapovednik generally with a shortage of staff. In order to reduce their dependence on State funding, in time of war, the remaining zapovednik staff attempted to be self-sufficient, so that a good deal of time was taken up with growing their own food.

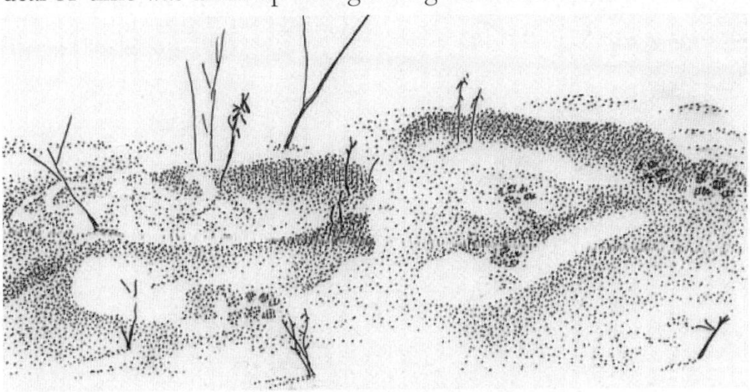

More tiger spoor: the 'sphinx position'

Tigers lie in this position when quietly studying an area, then sometimes stalk and eventually attack prey. The two front legs are directed forwards, parallel to each other, forming the two depressions in the snow on the left of the diagram. The back legs are somewhat splayed, and form the depressions on the right. The pug marks were made after the tiger got up (perhaps by the same tiger, perhaps by another). Modified from Matyushkin (1998).

On 13th May 1943 Kaplanov set off with a zapovednik guard to pay staff at a distant guard post. On the way he changed his plans, left his horse on the path, and walked up into the hills alone to inspect some salt licks. On 16th May the guard returned to the coast and reported that Kaplanov had not reached the guard post. Lida organised a search. Only after two weeks was his body found - deep in the forest, under two recently felled birch trees. He had been shot. They brought his body back, and he was buried at Valentin Bay.[22]

In a letter written in June 1936 he had declared that 'ahead of me lies the real work - to subjugate the Manchurian bear and the Lord of the Mountains and Forest; and I shall not give up until I have completed the task or have left my bones in the Sikhote-Alin Mountains'.[23]

At the insistence of Moscow scientists an inspector was sent to the reserve. After his investigation, he confidently reported that Kaplanov had caught a local poacher called Zuyev hiding the meat of a black bear he had shot the previous day; Zuyev saw Kaplanov first, shot him in the back, then turned him over and shot him in the face, before firing twice more. The alleged culprit was punished appropriately; but he was adamant, on returning home from prison, that he had not in fact committed the crime.

There are reasons for doubting the truth of the official version of events. Although Zuyev admitted to being a poacher, he probably did not murder Kaplanov. He may well have been charged and convicted to get others off the hook - and in particular some soldiers at a military guard post not far from the scene of the crime. The truth will probably never be known.[24]

Chapter 8

MIONU AND THE MEKA ROCKS[1]

Below the pass the River Sadomabirani[2] starts life as a small stream flowing west for five kilometres, before joining a similar one from the north-east. At the junction stood a yurt, with an Udege family busy around it. They were loading their sledge, evidently about to move away for good.

I asked who they were and where they were off to. The man told me that he was Mionu of the Kimunku clan and that they were from the River Khor. They had come here to trap sable. However, tigers had taken nearly all his dogs, and so he had decided to give way to the tigers and return to the Khor. I told him that I wanted to spend the night here, and asked him to stay with us until the next day, to show us where the striped beasts driving him away were most often found.

When his wife heard that he had agreed she began, silently and with no hint of annoyance, unloading the sledge and dragging everything back into the yurt. When the Cossacks arrived it was time to set up our camp. Everyone set to good-naturedly: some pitched the tent and got the stove going, some cut firewood, others collected herbage to serve as bedding, and Maruchin cooked supper. When it began to get dark I ordered all the dogs to be brought into camp and fires to be kept going all night.

Meanwhile Kosyakov and I visited the yurt. Mionu was about thirty-eight years old, quite short, and poorly dressed. His sun-burnt and weather-beaten face and calloused hands indicated what he had to endure to provide the essentials of life. In his wife's eyes I saw only resignation.

I began to ask Mionu about the river. He told me that all the right-bank tributaries of the upper Mukhen are considered good hunting country. The hills between the River Nefiktsa and Meka had been extensively eroded, resulting in many rocks with fantastic shapes. The area had more tigers than anywhere else in the district. Wherever they wander off to, they always come back. Up there they produce their cubs. All hunters know that, and so no-one wants to trap sable near the forbidden rocks.

The tiger that had taken Mionu's dogs had come from these rocks. It was an audacious animal. It began visiting the Udege family at night, but now was in the habit of wandering around in broad daylight. Only that same morning, a few hours before we arrived, it snatched yet another dog - the fifth. When Mionu shouted at the tiger, it bristled, bared its teeth, roared, and started swishing its tail back and forth. From this Mionu understood that he and his family were in mortal danger. So he resolved to return to the Khor, leaving behind a puppy as a sacrifice to the lord of the forest.

It was dusk when I left the yurt and returned to our camp. After supper I went back to the yurt. The children were already asleep, the woman was cooking their supper, and Mionu himself was repairing ski straps. I sat down by the fire and began asking him about the ill-reputed Meka Rocks. For a while he remained silent, and I wondered whether he perhaps did not want to talk about them, in case I asked him to take me there.

At last he answered, 'It - very bad place. Our people never go there.'

Then he explained. The Meka Rocks are unapproachable. They are where the devil Onku lives. There the wind is always blowing. When people go past someone is always whistling high above, and sometimes voices are heard, the sound of skis, the thump of axes. This is why the cliffs are called Onku Jugdyni, which means 'dwelling of the bad spirit Onku', or Amba Jugdyni, because there are always many tigers. The tigers are Onku's dogs; they obey his orders. Mionu talked at length about the terrors of the forest, which stimulated my curiosity. After supper his wife lay down to sleep, but we stayed sitting by the fire and continued talking amicably.

Gradually the topic of conversation changed. Mionu asked me about the life of the Udege down on the coast, and reminisced about some of his old acquaintances. Somehow our talk dwindled away. Memories flooded back and I recalled my long journey across Siberia, when I came to the Russian Far East. Mionu too was lost in thought, and stared silently into the fire.

Suddenly he started. Terror was written on his face. I saw him go pale, and hold his forehead with his shaking hand.

'What is it?' I asked anxiously.

'Kuti-Mafa close by,' he whispered, hardly audible.

I listened, but my ear caught not the slightest rustle. Then I took my rifle and went out.

It was a moonlit night. The moon, fuller and brighter than the night before, had just gone behind a cloud drifting from the north-west. Sometimes it appeared just for a moment, seeming to be floating in the opposite direction past the static clouds. The weak starlight, the pure frosty air and the silent flow of the water past an ice-free patch of the river - everything pointed to undisturbed peace in the forest. The river looked like a broad white boulevard, lined with tall old spruce trees and disappearing downstream into the forest. The tent was dark, from which I concluded that everyone was already asleep. Thin columns of smoke rose vertically from the dying fires. I could smell the ashes.

I looked around and listened, but the silence was complete. The perfect silence aroused my suspicions. Beyond the river the mute forest was a moonlit wall of trees. It was almost beyond belief that nature could be so quiet. It seemed as though the whole world was in a deep slumber.

I went back into the yurt and told Mionu that I had seen nothing untoward. He said that although I saw and heard nothing, the tiger was still close by: everyone knows that if a person suddenly takes fright without any obvious cause, it means that some fearsome beast has come close to their home.

I suggested that we could fire into the air once or twice, or throw some smouldering embers into the forest. Mionu said that would not help: the tiger would go off for a while, but come back later, and then things would be worse still.

Next day I was not intending to wake my companions early. However, when I began dressing, Glegola asked to come with me. Trying to make no noise we picked up our rifles and tip-toed out of the tent. The day promised to be sunny and cold. Silver-white feathery clouds stretched across the sky at a great height. It seemed as though the cold had congealed the air so that it no longer stirred. Frozen trees in the forest cracked with a ringing sound. Smoke from the fires formed a layer like mist, hanging above the ground.

Mionu was on his feet. His wife cooked breakfast while he cleaned his weapon and their children played with some lynx bones. During the meal he said he would guide us only to where the tigers are normally found, and then he would come back and, without waiting for us, set off for the River Khor. I knew there would be no point in trying to change his mind, and therefore reassured him that I had not forgotten my promise and would not detain him up beyond midday.

When everything was ready we put on our skis and followed our guide. He took us along the steep bank of the river through ancient forest. In many places the slope had collapsed, exposing the trees' roots. One spruce had fallen over, in the process creating a mound of soil. Here, on a snowdrift, was a well trampled path.

'No human track: all tiger,' said Mionu, pointing to the path.

It was clear that the tiger came this way frequently. The path was firm and the pug marks shone as if they had been imprinted on wet snow which had then frozen.

It would have been easy to set up a rifle or cross-bow trap here to wound the tiger severely, but Mionu had decided to cede to the tiger not only the hunting ground but all his remaining dogs.

At this point we began to climb. The snow under the trees was not deep, and so all the forest debris cast shadows which varied from pink to blue, depending on how high the sun had risen in the sky. The tiger path also showed up blue, until it split into several tracks. One was quite fresh. Mionu had been quite right: during the night the tiger had evidently visited the yurt. I noted that the spoor was not all of uniform size. Mionu had been visited independently by not just one tiger, but two, since the tracks did not coincide. They were often heading in opposite directions. It meant that two tigers were competing for the dogs. We took the fresh track and followed it for about a verst.[3] It headed up the mountain, leading us into forest debris and thickets of young spruce that were difficult to force a way through.

Here we found a very fresh lair. The tiger had been lying on its side with it head facing up. The snow retained a perfect impression, including the head and ears, neck, body, and extended legs. Its tail had gently moved about, sweeping snow to the side. We could read in the snow the sequence of events: something had suddenly startled it, it jumped to its feet, stood still for a while, and then ran off a short way before standing once more. We were probably the cause. Despite all our precautions the sound of the skis must have carried a long way in the still frosty air.

Mionu halted and said he would go no further. He advised us to take great care, since the tiger knew it was being pursued. He turned back and we continued tracking.

The tiger had made a long jump in the direction of the river. Now the track crossed itself not just once but twice, and it became clear that the tiger intended to defend itself and attack from the side. Accordingly I decided to abandon this dangerous game, and return directly to our camp. We soon found Mionu's ski track. For some reason he had chosen a new route, straight downhill through pathless forest. Only Marunich was in camp when we arrived. He said that the others had gone along the river hunting.

Around the yurt the family was hastily loading the sledge. I went up to Mionu. He was extremely agitated. Apparently he had almost paid with his life for being our guide on the tiger hunt. On the way back it turned out that Kuti-Mafa, going back on its tracks, arrived near the path we took on the way up. Hearing Mionu's approach its anger was roused. Mionu spotted it lying in wait on a fallen tree. There and then he turned off the track and descended as fast as he could into the Sadomabirani valley. I remarked that he had missed a chance to shoot it, but Mionu was determined to yield to the tiger.

When the sledge was loaded Mionu harnessed two adult dogs to it, tethered a puppy to a tree, and set off along our path up the Sadomabirani. His wife pushed the sledge from behind with a pole, and the children skied alongside. The puppy, sacrificed by Mionu to the tiger, pricked its ears and then, turning to face the tree, pulled with all its might at its tether in an attempt to get its head out of the noose.

Marunich went up to the puppy, released it, took it to our tent and fed it on *kasha*[4]. Somehow or other the puppy immediately got used to the new situation, and in half an hour was already trying to make friends with one of our dogs - though the latter did not take kindly to the puppy's familiarity: for a long time it growled and bared its teeth.

Not long before dusk the men returned from their hunting. They had all seen tiger spoor. Glegola had shot a musk deer.[5] It was a timely success, since we were getting fed up with *kasha*, and everyone hankered after meat. The temperature was falling rapidly. It was very cosy in the tent, with the red-hot stove giving off plenty of heat and the kettle boiling merrily on top. The Cossacks adjusted the straps on their skis and related what had happened during the day, laughing, and recalling Marunich's disastrous descent from the pass some days earlier when we had come into the valley. I wrote up my diary. When the puppy found its way into the tent the men stroked it and pulled its ears while it larked about, rolled on its back, and playfully bit their hands.

Since it was such a dangerous place, we decided to draw up a rota of guards for the camp. Once during the night there was a lot of barking from the dogs, so we brought them all into the tent and lit a big fire outside.

Marunich woke me just as dawn was breaking. I made the meteorological observations and prepared the plane-table for surveying. When breakfast was ready we roused the men. It took us an hour at most to get ready. Each knew what he had to do, and so taking down the tent, dismantling the stove, loading the sledges and harnessing the dogs all went like clockwork. When the dogs were harnessed the smokers in the party, as always, dashed over to the fire to have a smoke 'for the road'. Then, taking up the hauling straps and shouting to the dogs, the riflemen and Cossacks set off in file. About one hundred and fifty paces from the camp I glanced back and saw a thin column of smoke rising from the fire.

Starting the day's surveying always takes rather more time than continuing it on the journey. The pedometer needs calibrating, back bearings on the last section of the previous journey need to be taken, other bearings are needed on surrounding peaks, and so on. Finishing all that, I suddenly remembered that I had left the barometer in camp. There was nothing for it but to return. The men were well ahead by this time so I did not call them, expecting to catch them up at the next pass we had to cross.

Approaching the campsite I thought I saw something large and yellow disappearing into the bushes. I stopped and listened, but seeing nothing suspicious I decided I had imagined it. With no time to lose I went over to the tree the barometer was hanging on and took it down. To hang it around my neck I took my hat off, but then dropped it. Bending down to pick it up I saw large tiger spoor in the snow. Realising that I was alone and in the presence of such a fearful and audacious beast filled me with alarm. I hastily thrust the barometer into a pocket and put on my hat, looking around all the while. Then I set off back along the sledge track. I kept going steadily without running, but with my index finger on the trigger of my rifle. Soon I quickened my pace and in a few minutes reached the first bend in the river.

The valley widened out. In the distance near the next bend I spotted the others. My fear subsided, and I calmly continued with the surveying.

Map 5. World distribution of tigers

The dashed line with inward-pointing bars indicates a reconstructed former range of tigers when it was still continuous. Within it the hatched patches show the approximate present distribution. The dots are isolated records (1800-1950) of tigers to the north of the reconstructed range. (Modified from Kucherenko, 1985, p.42.)

Chapter 9

DISTINCTIVENESS, RANGE & HABITAT

The fossil history of the tiger *Panthera tigris* goes back some two million years. The most complete early tiger-like skull was discovered in China and named *Felis palaeosinensis*.[1] Fossils, also about two million years old, have been found in Java. Indeed the fossil record from Siberia, China, India, Java, Japan and the Caucasus broadly corresponds to the tigers' recent range, and indicates that tigers had a wide distribution in East Asia some 1.6-1.8 million years ago.[2] The evolutionary history of the tiger was probably closely linked with that of the larger members of the deer and pig families, whosc expansion provided opportunities for the powerful predator, and was also affected by changes in distribution during a succession of ice ages.[3]

The various populations of tigers became adapted to a great variety of environmental conditions through natural selection.[4] For instance colour variation in mammals is often influenced by habitat and climate, especially temperature and humidity:[5] tigers from the north, such as Amur tigers, tend to be paler in colour; those from the south, for example Sumatra, are generally darker. Representatives of a particular species can also show variation in size because of the differences in the prey they take: if the prey is large, the tiger must be too. Moreover tigers may be large simply if they are never short of food while they are growing.[6] Another tendency is for members of a species in the colder parts of its range to be larger - because this reduces the surface-area-to-volume ratio and so helps to conserve heat.[7]

Until comparatively recently eight subspecies of tigers were recognised in Asia.[8] Of these, five are still in existence - the Indian, Indo-Chinese, South-China, Amur and Sumatran. Of these the South China tiger, a denizen of sub-tropical and temperate forests, is almost extinct. The Bali, Java and Caspian tigers are definitely extinct.

Recent research suggests that the varieties of tiger in Asia should be regarded as populations adapted to different habitats rather than as discrete subspecies. Of the five surviving tiger races or populations most experts now consider that the Amur, South China, Indian, and Indo-Chinese tigers are not sufficiently distinct to warrant being called subspecies. Sumatran tigers, however, isolated on an island probably for 10,000-12,000 years, are sufficiently distinct to be considered a full subspecies *Panthera tigris sumatrae*.[9]

Distinctiveness of the Amur tiger

The question arises: if the Amur tiger[10] is not clearly different from tigers in other areas, is it worth conserving? Even if the kinds of tigers are now regarded as regional and intergrading populations adapted to different environmental conditions, it can be argued that they should be preserved just for that reason. This would support an approach to tiger conservation based on ecosystems rather than on subspecies:[11] protecting the habitat helps conserve the tiger, not only by providing space and prey for the tigers' survival, but by maintaining the varying ecological conditions which produced the variety of tigers we have today or have had until recently. The Amur tiger may be regarded as a form specially adapted to the temperate conditions in the north of the species' range, where winters are long and snowy. Even if the Amur tiger can no longer be considered a distinct subspecies, its characteristics are still worth preserving in their original form.[12]

Amur tigers are massively proportioned animals, with a heavy bone structure, a huge skull and large canine teeth.[13] In addition their long thick silky winter fur gives them a robust appearance. This is one of their most striking features, and although the winter coats of Indian tigers are heavier than their summer ones no other tigers have fur as long as the Amur tiger's. Adult male Amur tigers have much larger heads and bodies than the females, and the impressive mane-like ruff around the throat is especially well developed. These tigers also put on a layer of fat for insulation during the winter. In summer, after moulting, the animal has a shorter glossy coat.

Amur tiger

Among the largest of the tigers, it is usually an ochre-yellow colour with black stripes and white on the face and underparts. The winter coat is luxuriant and thick, and the long tail is bushy and hardly tapers. Around the massive head is a ruff of extra-long fur. There is a white spot on the back of the ear.

The Amur tiger's mane or ruff is well developed, and is striking when the animal snarls.

The typical background colour of the Amur tigers' winter coat tends to be rather lighter than that of most other tigers. It is usually an impressive pale golden ochre-yellow with an admixture of reddish-rust, and may shade to a richer orange colour in some individuals. Its summer coat may be of a brighter or more orange hue. Compared with most other tigers the Amur tiger has larger areas of white on the face, chest and underside, where it may extend quite far up the flanks.[14]

Genetic tests have shown that the wild Amur tigers in the southern Russian Far East, descended from the very few individuals that survived in the 1930s, are all very closely related.[15] There is however considerable individual variation in stripe pattern: stripes may be broad and quite widely separated, narrow and close-set, or narrow and well separated. The white, black and ochre colouring matches the background colour of snow, shadows, and yellowing dead oak leaves in the scrub oak woodland where they are often found. Their striped pattern breaks up the outline of the animal.

A series of Amur tiger pelts in Moscow University Zoological Museum indicates that specimens from the northernmost part of the tigers' range tend to have lighter and yellower background colouring and broader, more widely separated stripes than those from farther south. Formerly it was believed that there are two races of tigers in northeast Asia - the Amur tiger in the more northerly part (Khabarovsk Region and the Bureya Mountains), and the Korean tiger[16] in southern Ussuriland, Manchuria and Korea. The Korean tiger was described as brighter or darker orange in colour, more heavily striped, and without such a long coat as the more northerly tigers.[17] However, some Amur tigers from the north of their range also show these latter characteristics, with some being quite a deep orange.[18] In earlier times, when the ranges of the Amur and the smaller South China tiger farther south overlapped, intermediate forms would have existed.

The Amur tigers' reputation as the world's largest cat has often been emphasised. One source gives typical weights of 180-306 kg for adult males and overall lengths including the tail of 2.7-3.3 metres.[19] The Russian tiger expert Sysoyev published a weight of 384 kg for a specimen collected in the Sikhote-Alin in 1950,[20] and Baykov recorded tigers shot in Manchuria of 390 kg and even 400 kg! [21] But at that time there was no scientific verification in the field.

There may be several reasons why such enormous specimens have been reported in the past but not recently. It may be because of the selective shooting of larger older specimens.[22] Another explanation depends on the size of individual large predators being related to the food supply - and prey was certainly much more abundant in Manchuria and the Russian Far East in the early twentieth century when these huge tigers were reported. Or, quite simply, when hunting was not so intense, individual tigers may have lived longer and so had a chance to grow larger. However, other experts believe that such reported dimensions were exaggerated.[23]

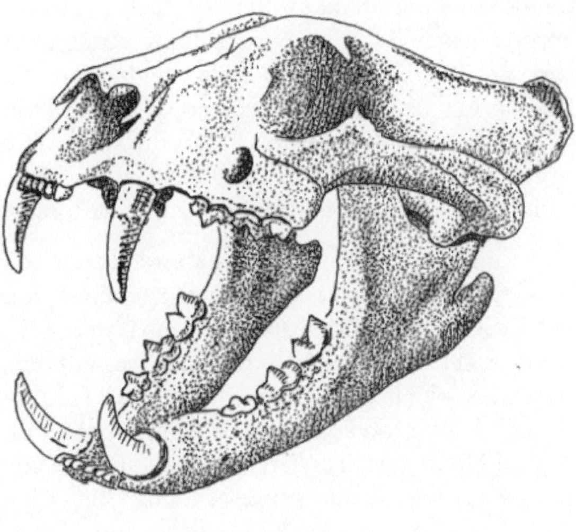

A tiger's teeth are divided into three sets - in front, the small upper and lower incisors, twelve in all; then the massive stabbing canines; and finaly the very sharp molars and premolars, used for slicing up prey. There are no chewing teeth. (Drawn from a specimen in the Royal Museum, Edinburgh.)

A study of the carcasses of eight wild Amur tigers from the Russian Far East, which had died of various causes, showed the heaviest to be a well fed male, about 4 years old, that weighed 193 kg.[24] The Amur Tiger Project, which has operated from Sikhote-Alin Zapovednik since 1992, has found that 215 kg seems to be the heaviest wild male Amur tiger that they have been able to verify, albeit from a limited number of specimens. Although it is certainly among the largest of tigers, comparable to those from northern India and Nepal, the claim that the Amur tiger is exceptionally has possibly become embedded in the literature without any real justification. Records of immense Amur tigers may be of captive specimens.

In addition to the reason already given for protecting it, this impressive animal is important as a flagship for the conservation of a unique ecosystem of mixed deciduous/coniferous forests with exceptional biological diversity. The Amur tiger is also important as a cultural icon - a magnificent symbol of the heritage of the indigenous people, Russians, Chinese and Koreans, which gives special significance to the region.[25]

Range (see maps on pp.x & 14)

Although their main stronghold was in the temperate forests of the southern Russian Far East, northeast China and northern Korea, Amur tigers were found in former times over a much greater area of northeast Asia. Their present northernmost limit is roughly along the Amur River but there is evidence of tigers wandering much further north. In the museum at Yakutsk there is a mounted specimen shot in the area in 1905.[26] This tiger may have been an occasional visitor, or was possibly driven north by the Russo-Japanese War of 1904-05.

Tigers were permanently resident in the mixed conifer-broadleaf forests on the left bank of the Amur, especially the Bureya River basin, as far as approximately 52°N, with individuals wandering westward into the area east of Lake Baykal. There is also evidence of tigers inhabiting suitable habitats in the extreme eastern part of Mongolia in the region of Khalkin Gol in the early 19th century.[27] To the south, in earlier centuries, Amur tigers may have reached latitudes just south of Beijing, in Hebei Province, which was originally covered in temperate forests.[28]

China and Korea

Until the nineteenth century Amur tigers were found over a large part of Northeast China and Korea. In China, the most northerly limit of the tigers' range would have been in the Greater Khingan Mountains[29] in the northwest of Heilongjiang Province; they form a natural divide between the dry steppe and semi-deserts of Mongolia and the moister steppes of central Manchuria. The larch and pine forests constitute one of the few remaining extensive forest reserves in China - home to the hunting and reindeer-breeding Evenki or Orochen tribe.[30] When the western stretch of the Chinese Eastern Railway was being built from Hailar to Tsitsihar through the mountains in the 1890s, it was recorded that a regiment of Cossacks was employed as guards to shoot the tigers that were carrying off the workmen. The animals even entered the huts of Chinese and Russian settlers and attacked the inmates. Tigers and leopards have not been reliably recorded from any part of the Greater Khingan since the 1960s.[31] Hunters are very active in the area, taking a heavy toll of elk, roe deer and wild pig. Railways and roads for the exploitation of the great stands of softwood are penetrating ever more deeply into the highlands, so that the rich natural heritage of the region may be lost before it has even been properly studied.

Further southeast along the right bank of the Amur, the Lesser Khingan Mountains[32] continue for nearly 500 km, linked to the Greater Khingan in the north-west so that the whole Khingan system forms an inverted L shape. Great forests of larch, ash, pine, spruce and fir have survived until recently, and to the east there are more diverse mixed conifer-broadleaf forests. The region has been rapidly opened up to logging, the forest cover has decreased by 40-60% from the 1950s to the 1990s, and the human population of the region has increased more than 16 times in the same period.[33] The remote and sparsely populated Lesser Khingans used to be rich in wildlife; once common there, tigers have not been recorded since the 1970s.

The rich forests of the southern part of the Amur tigers' range in southeast Manchuria may have provided optimum conditions for the Amur tiger's existence with abundant prey, vegetation cover and water.[34] Along the Russian frontier - on the left bank of the Ussuri River and along the Korean border - tigers were numerous in the nineteenth century. These magnificent forests - the *shou-hai* or 'forest sea' - especially near the border, west of Lake Khanka, were the setting for Baykov's writings on the life of the Manchurian tiger.

The Changbaishan runs northeast to southwest as a series of volcanic ranges for about 1,000 km along the border between China and North Korea. In the reign of the Qing Emperor Kang Xi, in the eighteenth century, officials reported on the rich natural resources and extensive well-preserved forests of the region. But they have been drastically reduced; a 14,000 sq. km area of primeval forest in 1870 had contracted to only 573 sq. km in 1980.[35] Tigers probably died out in the Changbaishan in the 1980s, although sporadic sightings have been reported from this region.[31] In 1998 four farmers in the Changbaishan were each granted 1,000 yuan (US$120) for the loss of three cows to tigers. A 1996 law provided that compensation would be paid to anyone whose crops or domestic animals are damaged by wild animals.[36]

The Changbaishan Mountains and Nature Reserve in Jilin Province are adjacent to the Paektu San Mountains and reserve in North Korea. In the mid 19th century, the tiger was found in the rich forests and river valleys of the mountainous peninsula of Korea, extending to the southernmost parts. It was reported in 1870 how 'wolves and tigers [abound] in the north: their skins form part of the tribute to China'.[37] Tigers are now certainly extinct in South Korea and although there have been very few confirmed sightings in North Korea since the 1970s the authorities have stated that about ten may remain in the forests along the Chinese frontier.

Historical sources indicate that in the past virtually every forested area of Heilongjiang and Jilin Provinces had Amur tigers. Although by the 1930s the tiger was rare in Manchuria, it was estimated in the 1950s that there were still some 200 in the area.[38] Since then the number has continuously declined, though the Chinese say that there are still a few resident in the Wanda Shan in eastern Heilongjiang Province, where the Qixinglazi Nature Reserve was set up in 1980. With inadequate protection even within reserves the range of the Amur tiger has retreated to the east, and the population has almost been extirpated. By 1975, in Heilongjiang Province 81 tigers survived in three mountain areas in the east of the province; by 1991 only about 12 remained in the same area.[39] In the winter of 1999-2000 a team of Chinese, Russian and US scientists investigated the remaining tigers of eastern Heilongjiang Province: the results suggested that 5-6 were present in the Wandashan and Zhangguangcailing forest districts - all but one in the former.[40]

1998 was the first year in which a joint survey of tigers in Jilin Province was carried out by a team of Chinese, Russian and American tiger specialists.[41] Evidence was found of 4-6 tigers and 4-7 leopards. Without exception they were close to the Russian border. In the 1996 tiger census an adult male and female tiger with two well-grown cubs were found to include within their territories areas on both sides of the frontier - despite the border security fence! The 1998 Jilin survey found rather few tracks of suitable tiger and leopard prey, but plenty of wire snares together with their victims.[42]

Tiger Conservation in China

Extensive tracts of forest lands remain in parts of Jilin Province. They are well managed by the Ministry of Forestry, and at least retain the capacity to harbour good populations of prey animals, as well as tigers and leopards. The 1998 survey report recommended the development of a recovery plan for tigers and leopards focusing on Hunchun County and aiming to re-establish a breeding population of the big cats in Jilin Province. The hunting ban is being enforced, illegally held firearms are being collected, and effective action is being taken against the setting of snares.[43]

In northern Hunchun a new nature reserve was set up in 2001 in excellent tiger and leopard habitat. It is situated along the borders between southern Heilongjiang Province and Russia. Some part of this could be included in a proposed Panthera Big Cat International Park that would include the Borisovkoye Plateau Zakaznik[44] on the Russian side and a new nature reserve created in the Laoyeling range in southern Heilongjiang in 2001. The Hunchun Reserve is patrolled by fifteen rangers, and in 2002 more than 5,000 snares were collected there. In February 2003 for the first time a tiger was photographed at night by a camera trap in the reserve. This shows that tigers cross the frontier from the Russian Far East; but as it was feeding on a horse taken from a local peasant, there is concern over the insufficient natural prey on the Chinese side of the state border.

In southern Hunchun County along the Tumen River, a 10 km strip of forested mountains provides a potential ecological corridor between Russia and North Korea. Evidence gathered during the 1998 survey along the Chinese-Russian frontier raised hopes that tigers from Russia are still able to disperse into North Korea through Chinese territory. Fragmentation of existing habitat and elimination of this potential linkage are threatened by proposed industrial deve-lopment of the region. Therefore joint action will be necessary to preserve the integrity of the tiger and leopard populations, with the

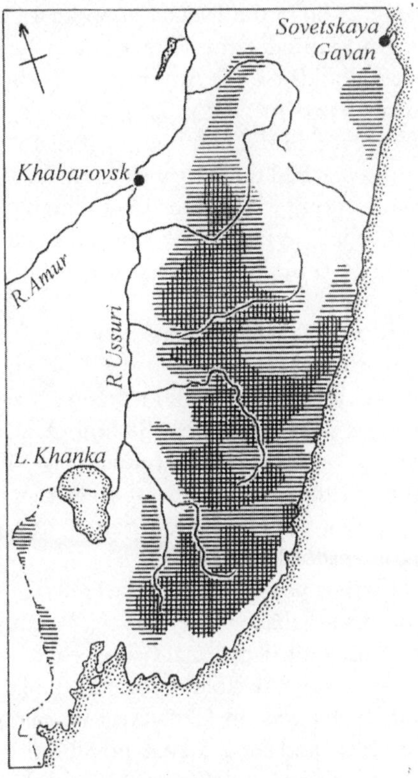

Map 6. Range of the tiger in the southern Russian Far East

Based on the situation in 1983. Single hatching shows areas normally inhabited by tigers; double hatching indicates areas of high density. There are also many records of tigers from surrounding areas (not hatched). There has been some contraction since 1983, particularly west of Lake Khanka, and in the isolated area near Sovetskaya Gavan. (After Kucherenko 1985.)

ultimate objective of forming an international reserve involving all three countries.[45]

Russian Far East

Despite the former importance of China for the Amur tiger, in the 21st century the Russian Far East provides the best chances for its survival. Although it has almost disappeared from Northeast China and North Korea its range in the Russian Far East along the approximately 1,130 km of the Sikhote-Alin has remained essentially intact. In 2003 some 400-450 tigers were surviving in the Russian Far East - compared with only 25-30 in the 1940s but about 1,000 in the mid 19th century.

Undoubtedly the Sikhote-Alin has always been one of the main strongholds of the Amur tiger. Its diverse mixed conifer and broadleaf forests provide the most favourable habitat for its main prey species. However, in the extreme northeast of their range the tigers in former times reached the boreal forests of the lower Amur river, and even as far as its mouth. In winter they may have occasionally crossed the ice to Sakhalin. Certainly they reached Lake Kizi on the lower Amur, and the rivers Gorin and Gur that mark the northern limit of the Manchurian forests.[46]

Along the Pacific Coast they do not seem to have been found regularly much further north than the Tumnin River and Sovetskaya Gavan. As the dark-coniferous Okhotsk forests replace the richer Manchurian mixed forests in the northern Sikhote-Alin, the depth of the snow cover is greater and there are fewer wild pigs. Tigers, although present, are rare at this northernmost limit of their range on the coast. However, further inland across the mountains they regularly visit the environs of Komsomolsk-on-Amur.

Farther northwest tigers were once common in the mixed forests of the Biro Bizhan and the Bureya Mountains as far west as 129°E, but now they are only occasional visitors to this region. There have been, however, a number of reliable reports in the late 1990s of tigers occurring quite far to the west in Chita Region, east of Lake Baykal. Tigers are great travellers and these animals possibly dispersed westward from Khabarovsk Region because of forest fires in 1998-99, rather than having crossed the Amur from China.

The greatest density of the Amur tiger in their stronghold of the Sikhote-Alin is in the southern and central parts where the forests are more diverse. Tigers inhabit the eastern macroslope (43-44°N), especially in the region of Lazovski Zapovednik; both sides of the main watershed between 45-46°N, including the Sikhote-Alin Zapovednik; and, further north, the Bikin and Khor valleys. These areas still have significant populations of hoofed mammals and provide favourable habitats for the tigers as well as having few people and good levels of protection.

In the 1990s poaching was severe in the southern parts of Primore accessible to large urban centres such as Vladivostok, Ussuriysk and Nakhodka. Future expansion of human settlement is likely to follow the river valleys, and the Khabarovsk-Nakhodka main road will fragment the tiger's domain still further. However, the Amur tiger has retained about 60% of its continuous range in the Russian Far East, which is probably better than in any other part of Asia.[46]

Habitat

Siberia conjures up images of desolate snowy wastes, and the Amur or Siberian tiger has even been said to live 'far beyond the Arctic Circle in the snows of Siberia'.[47] On the contrary, the Amur tiger's home country is the Sikhote-Alin and adjacent areas - a very distinctive and un-Arctic region of temperate mixed broadleaf-conifer forests. The northern limit of the tiger's range on the Amur River is on about the same latitude as southern England, approximately 52°N, although the climates are not at all the same.

Chapter 3 described how the Sikhote-Alin forests comprise two separate floras - the Manchurian and Okhotsk.[48] According to native legend two sacred birds flying from north and south collided in Primore scattering their burden of seeds, and this is how the diverse flora of a temperate rain forest came to be intermingled with that of the fir-spruce taiga.[49] Northern mammals such as the lynx, wolverine, wolf, elk and brown bear are found alongside species more closely associated with the Indian fauna - such as the Asiatic black bear, yellow-throated marten, Far-East forest cat and goral. Smaller carnivores include the raccoon dog, badger, sable, Siberian weasel, stoat, weasel, river otter and red fox. There are wild pigs as well as *izyubr*, roe, sika and musk deer.

Large carnivores such as the Amur tiger and Far Eastern leopard are dependent on these large hoofed mammals as prey. With a shorter growing season than in southern Asia and more limited fodder, hoofed mammals are more spread out and so Amur tigers need much larger feeding ranges than those in India. The northern tigers have probably always been few and far between.

Before the intensive impact of humans the mountains of the southern Russian Far East and Manchuria were covered with unbroken mixed forests of great biological diversity. The vast forested belt of the Sikhote-Alin ranges with their boxed-in valleys had almost no natural openings below the tree line and this feature, clearly discernible from the air, has acted in the past as a natural protection for the Amur tiger.

In some areas of the Sikhote-Alin, primary forests have been modified by logging and fires into secondary forests and thinned out woodland to form a mosaic with open meadows and lakes. These areas are very suitable for *izyubr* and roe deer, and thus also for tigers. But tigers scarcely visit the dark spruce-fir forests at high altitudes, where there is little suitable prey.

The main tiger strongholds are Sikhote-Alin and Lazovski zapovedniks. Writing about the former, the well known Russian tiger expert Evgeni Matyushkin noted that 'particularly favourable habitat for tigers are the pine-broadleaf forests in the valleys and on the mountain slopes in the middle reaches of the rivers on the eastern macroslope, where there is a rugged relief often with cliffs and rock outcrops, for instance on the River Serebryanka', which runs down to Terney.[50] In Lazovski Zapovednik, where conifers are fewer and the vegetation more southern and diverse in character, tigers occur almost everywhere. More than half of Lazovski Zapovednik is covered with secondary oak forests and the density of the tiger population is highest in those areas near the coast where the forest has been altered most. With dense undergrowth and richer feeding, it is most favourable for the sika, red and roe deer.

Although the focus of the Amur tiger's domain has been the biologically diverse temperate mixed broadleaf-conifer forests, in former times it was found in a variety of habitats - in fact wherever there was suitable vegetation cover, water and sufficient prey. It was reported from the grassy marshlands and reedbeds along the main river valleys including the Amur and Ussuri, and from the long grass monsoon steppe of central and northeast Manchuria. Although providing some of the best tiger habitat, river valleys were the first to be settled and cultivated by humans.

It was observed in the mid 19th century that tiger tracks were frequent in the reed beds and alluvial grasslands around Lake Khanka.[51] This is one of the largest freshwater lakes in Asia - covering 4,400 sq.km but nowhere more than 10 m deep. The southern shoreline of this unique wetland habitat is included in Khanka Zapovednik. After a 20 year interval when tigers were not recorded from the area around the lake, a three year-old-male tiger was found drowned after a violent storm had flooded the shore.[52] According to the director of the zapovednik Yuri Sushitski, the tiger had been living in the reed beds and preying on raccoon dogs, which are numerous in these marshes. Apparently it had attempted to swim to higher ground but had been overcome by the flood.

Chapter 10

TIGER NIGHTS [1]

Dense damp taiga. Enormous pines and massive hoary spruce trees pressing in on all sides. Under some gloomy overhanging rocks flickered our small campfire. We had come to this wilderness of trees and rocks at the Blue Mountain to hunt the king of the Manchurian mountain animals. With me was Sgt Major Dolgikh - a keen and courageous hunter.

Tigers had started disturbing the workers at the forest concessions to the south of Veyshakhe Station. A horse had been killed. We arrived the day before, and were now beginning to explore the district. We visited the cabin where the tiger had come the previous night, and drew up a plan of action. This cabin is situated at the foot of the rocky mountain, just at the edge of the forest. At night the ten work horses are tethered nearby, out in the open. In the cabin live up to twenty Chinese forest workers. Similar cabins are everywhere, in all the valleys, including below the Blue Mountain. The concession base with all the workshops and equipment was about three versts[2] to the north. From the station to the base stretched a 30-verst branch line.

Up until December there had been no sign of tigers anywhere near. It was said that they had moved south, following herds of wild pig. The pigs had left the area in search of food since neither the pine nor the other nut trees had produced a crop this year.

Now, around the settlement Rozhdestvo,[3] tigers had made an appearance near the railway. Their spoor had been noticed here and there, and eventually they had begun to pester the Chinese workers at the Skidelski forest concessions. The horse was killed on 10th January, and this prompted me to go to the Blue Mountain forest to try my luck.

At 10 o'clock in the evening, the Chinese told me, they heard a noise coming from where the horses were tethered - neighing of horses and growling from some other animal. Then there was the cracking of branches under the heavy footsteps of the predator as it left. Afraid to leave the cabin, the faint-hearted Chinese allowed the tiger to drag the horse about a hundred paces away and eat almost half of it. The tiger took the outermost horse, bit through the back of the head and the neck vertebrae, and then dragged the carcase downstream along the river. There, judging by the spoor, it stayed until morning and consumed the hind quarters and part of the back. It was a large tiger, old and experienced. The pug mark of its front paw was 4.5 *vershok*[4] across.

Six Chinese soldiers armed with Mauser rifles had gone after the tiger. However, coming across the spoor of several tigers in the mountains, the soldiers hurriedly returned.

Although more than a day had passed since the event, we decided to follow the tiger's track. Crossing the railway line, the animal had made off along the forest road up the mountain. By midday we were in clearings part-way up. Here, among almost impenetrable vegetation, we lost 'our' tiger among other tiger tracks. Some were quite fresh. Clearly, several tigers were wandering around. A female was on heat, and several males were following her.

The exhausting struggle to make headway along steep rocky slopes and sometimes in deep snow forced us to go down to a stream. Its water murmured under the stones at the bottom of a deep ravine. We had a quick bite to eat and made tea, thirstily drinking the fragrant liquid. From far away down below could be heard the thumps of axes and the crash as each forest giant was felled by ruthless hands. There, where a few days ago stood virgin forest, untouched by human hand, could now be heard the whistles of steam locomotives and the sounds of saws and axes. A railway had been extended to the foot of the magnificent and hitherto inaccessible Blue Mountain, and along it were being borne the mortal remains of the taiga giants - enormous trunks of pine, ash, spruce and walnut.

I was filled with sadness at such destruction of beauty in this magical primeval nature. In just a few years nothing would remain of the former Manchurian pine forests except memories. Where will the numerous herds go then - the wild pigs, *izyubr* and roe deer, and those regal predators the tigers? No doubt everything will be destroyed by human greed. But, for now, the lover of nature and hunting still has somewhere to roam. There are many other untouched virgin forests in Girinski Province, and still full of all kinds of game.

Our hunger and thirst satisfied, we once more set off along the tiger trail, climbing ever higher up the regal mountain, the immaculate snow on its majestic summit shining white against the blue of the infinite heavens. When the pine forest came to an end we entered a belt of fir and spruce. It became noticeably cooler, and the slightest breeze cut into our faces and hands. Here the animals' tracks separated, most descending the south slope of the mountain. We headed in the same direction, aiming to spend the night somewhere among the rocky outcrops in a pine grove. The sun had already disappeared behind the peaks of some distant mountains when we halted at the edge of the pine forest under some overhanging rocks.

Our camp fire crackled cheerfully below a massive old pine. Its dark branches swayed overhead as the flames flared up. Silently and hardly noticed, night descended on the mountains and forest. Stars shone through the dark network of branches of the taiga giants. The moon rose from behind a distant ridge and cast its pale phosphorescent rays into the depths of the dense forest, illuminating patches of snow under the trees, while everywhere else remained pitch black. The frost became more severe. Occasionally cracks could be heard from tree trunks and ice on forest streams.

The kettle was soon boiling. After drinking hot tea, we made beds for ourselves out of spruce branches. All around, silence reigned. Only now and then could be heard the trees, like the roar of a distant ocean. In the depths of the dark valley an eagle owl called: its loud 'oo-hoo' resounded clearly in the stillness of the night. The kettle was boiled and emptied several times. Warmed by the embers of the fire, turning first on one side and then on the other on our soft 'mattresses' of fragrant spruce, our tired limbs relaxed and we drifted off to sleep. The fairy-tale taiga night floated over the sleeping land and sent us magical dreams.

However, it was so cold that we could not sleep for long. The cold penetrated under our clothes so that now and then we had to turn our shivering side to the life-preserving fire. Waking up on one occasion because of the cold, I heard not far off the call of some kind of animal. It came from lower down on the steep slope opposite. At first I took it to be an Indian wild dog,[5] but soon another joined in, louder and more powerful. Now I realised that they were tigers. I woke up Dolgikh, who sat up by the fire and began to listen. Occasionally the animals' voices became louder and turned into a roar and thunderous coughing, but at other times they recalled the sounds made by fighting cats.

A tiger's call is something like the sound 'ye-o-oon', repeated once or rarely twice. The animals seemed to be coming closer, so we built up the fire and walked in the direction of the tigers, rifles in hand. We were expecting them to appear, but were mistaken: they did not come closer, and their voices instead died away in the stillness of the taiga. Judging by the number of calls and their relative strengths, there were at least four. One seemed to be a giant of an animal, roaring only rarely but very loudly and abruptly. When it stopped, the others could be heard - not so deep-throated and terrifying. Sleep no longer possible. For the rest of the night we sat by the fire, listening to the music of the primeval taiga.

The tigers' roaring lasted, off and on, for about an hour. To begin with it made a deep impression on us. We thought we were about to be attacked. Later however, when we had become used to it, we were not so alarmed - though we still kept hold of our loaded rifles. As well as the yelps and roars could be heard other noises - bushes and branches being broken where the animals were fighting.

It was an hour before sunrise when the tigers' calls gradually faded away, with just the occasional distant miaowing, disappearing in the depths of the forest. Briefly complete silence took over, but then the ancient taiga ever so quietly began whispering its own wild song.

As dawn began to break and the stars disappeared in the lightening sky, we carefully made our way, step by step and keeping as quiet as possible, to the site of the 'tigers' night'. On the side of the mountain, half a verst from our fire, we found fresh spoor of four tigers. In many places the snow was completely flattened. Where the animals had been lying down the snow had melted away altogether. Blood could be seen everywhere, and in one lair it formed a small pool, evidently coming from a wound in a left paw. It was clear that males had been fighting over a female. Judging by the tracks leading away from the battleground, three tigers had gone off along a trapper's path, across a pass and over into the Khami-khe-ra valley to the south: these were the two young males and the female. The spoor of the old male, with traces of blood, led us up the mountain through a ravine.

Old heavy tigers generally prefer to keep to lower levels in mountains, or else take to the high slopes - but always in places where rocks and cliffs emerge above the forest to provide a vantage point over the surroundings. There they may lie for days on end. Young tigers are encountered mainly in high rocky mountains, descending to valleys only to hunt, after which they return to the cliffs, where they rest in their favourite lairs in open places.

After proceeding along the tigers' path about three versts, and convincing ourselves that they were not going to stop soon, we turned back and began carefully climbing the mountain on the track of the wounded tiger. Its wound was evidently quite serious since there was plenty of blood along the way, and the tiger had lain down frequently and licked the injured paw.

The sun had climbed high in the blue sky and it felt quite warm in the bright blinding rays. The snow had noticeably melted in sunny places, refreezing to form a hard crust which crunched under our feet and seriously hampered our efforts to stalk the wary beast. Aware that it was being followed, it began circling around and backtracking through vegetation so thick that nothing could be seen fifty paces in front. It did not run away from us, but kept going slowly, about two hundred paces in front, all the while keeping us within earshot. Occasionally, when we stopped for a rest, it lay down in the snow facing us and licked its paw. So it led us on across the slope, sometimes taking us up onto rocky outcrops and steep slopes, at other times dropping down into ravines and valleys.

On one of the steep slopes we noticed fresh signs of wild pig feeding. We abandoned the tiger and followed the new tracks. In a clearing below rocks five pig were rooting. Approaching to within two hundred paces we took cover behind a fallen tree and opened fire. The pigs instantly made off, except for one wounded. It turned out to be a youngster weighing about five *pood*.[6] Disemboweling it and having a snack from what God had delivered to us, we covered our prey in snow and resumed tracking the tiger.[7] Imagine our surprise when we discovered from the spoor that the tiger had also been tracking the same wild pig! We could not help laughing heartily at ourselves and at the tiger's cunning and effrontery.

It is amazing that men and tiger - all three - were stalking the pig, though from opposite directions. We had approached them from above, while the tiger went around us and crept up behind rocks below them. If we hadn't been in such a hurry to open fire we'd have been able to watch an interesting spectacle as the tiger attacked the pigs. Of course he would also have been shot, and would have had to part with his splendid striped coat.

After our shots, it disappeared downslope in a series of leaps. In what remained of the day before the sun dipped behind the distant mountain ridges we continued tracking the cunning animal, but without success. We counted on giving it no peace and exhausting its patience, but it clearly sensed danger and refused to confront us.

The night closed in quickly. It soon became dark in the dense pine forest, and shadows filled the ravines. Sparkling stars shone in the heavens, and the mysterious peak of the Blue Mountain, illuminated by the silvery rays of the rising moon, stood out clearly, high above the surrounding dark forested hills.

At the foot of the mountain we came out onto a path which brought us to a track leading to a Chinese foresters' cabin. The small cosy cabin stood under an outcrop and on the bank of a taiga river. About 50 paces away six horses were tethered. Nearby were sledges and a simple harness.

The Chinese were already lying on the warm platforms[8] when we entered and greeted them with '*Ni lyao khao!*'. One after another, vying with each other and with shouts and gesticulations, they began telling us about how there were many tigers in the forest, and how they saw, this very day, a particularly large one - '*tatady-lyao-khu*'. Holding their hands about one and a half *arshin*[9] above the ground, they indicated its shoulder height. According to them its spoor was so large that a cap would not cover it. While listening to their excited descriptions, we drank tea and ate our way through some tasty Chinese rolls.

One of the Chinese was the foreman, and he was armed with a .375 calibre infantry rifle. He told us that, some days before, on his way back from work in the forest, he watched a tiger four hundred paces away. The animal came down from the ridge into the narrow valley and, spotting him, stopped with its tail swishing back and forth. It raised its head, which was broad and with a prominent forehead. The man decided to go no further, fearing an ambush, and fired twice into the air. The tiger disappeared into the undergrowth. He told us there were *izyubr* to the west of the cabin, among larch in the foothills.

Meanwhile night fell. One by one the Chinese lay down on the warm sleeping platform, covering themselves with dirty cotton blankets. One of them remained on guard outside, keeping a fire going right by the horses in order to scare off predators.

The only sounds disturbing the stillness of the night were the gentle snoring noises of the horses and the rustle of straw being ground up by their strong teeth. The guard, dressed in a Russian knee-length sheepskin coat and felt boots, was squatting by the fire, smoking a long pipe, and dozing, just occasionally opening his watery sleepy eyes.

In the cabin it was stuffy and smoky. Dolgikh was lying on the platform and seemed to be fast asleep, tired out by our exertions in the taiga, but I could not get to sleep. I went outside to get a breath of fresh air, and to admire the magical taiga night. All around, the pines formed a dense black wall. In the starry heights the horned moon cast its silvery rays onto the dark forest, the squalid cabin, and the white blanket of snow, sparkling with various colours. In the dark sky to the east towered a blue tent-like shape, stunning in its cliffs and snow. It seemed as though the magnificent summit of the regal mountain was alive and breathing, its mighty chest rising and falling, and the dark forest covering its steep flanks swaying. Its quiet breathing was heard in the rustle of dry leaves and the sound from the dark tops of the mighty firs and pines.

It was two o'clock in the morning. I went back in, where in the pale glimmer of the simple lamp - made from a bottle bearing the label '*Zubrovka Petra Smirnova*'[10] - I could only just make out the dark shapes of the sleeping Chinese. I felt sleepy and looked for a space on the platform where I could lie down.

Just then the guard ran headlong into the cabin shouting '*Lyao-khu!*' It took only a few seconds to seize my rifle and run outside. The frightened horses were pricking their ears and straining at their tethers. The Chinese hastily dressed and spilled out of the hut. The foreman cocked his rifle and was about to fire into the air, but I persuaded him not to, in the hope that the tiger would attack the horses and so give us a chance to bag it.

Accordingly the Chinese went back into the cabin to sleep, while just Dolgikh and I remained outside with two Chinese - the foreman with his rifle, and the guard. But the horses refused to calm down, pressing close to the comforting fire and flaring their nostrils when breathing in, and at the same time sensing the tiger's presence.

However intently we peered into the darkness of the forest we could see nothing. Evidently the animal was working its way around the cabin without showing itself. With rifles at the ready we walked along the path about 50 paces and stood under an old spruce in the shade of its thick branches. Meanwhile the foreman sat down by the water trough near the horses.

For about half an hour the taiga stillness was undisturbed. Gradually the horses calmed down and the guard began to hum a doleful and wild song. It seemed that the tiger had departed.

We were about to return to the cabin, since the cold was becoming unbearable, when suddenly the loud snort of a tiger came from right by the horses; it turned into an abrupt muffled roar. The horses began to shy away, trying to break free, the Chinese shouted at the tops of their voices, and the foreman would probably have fired into the air if we hadn't run up in time. Following the roar, a powerful cough-like sound came from the nearby ridge: it was another tiger.

Expecting the hungry predators to attack at any moment we took up position by the horses, using them as cover. The Chinese crowded around the door of the cabin, ready to retire if an attack came. The tiger on the ridge came down and approached the cabin, its voice growing louder as it came nearer. Soon the tigers were together. Their satisfied purring could be heard in the undergrowth, near the dark tops of some felled pines.

The horses, sensing the presence of humans, stopped pulling at their tethers, but continued neighing in alarm, shivering feverishly, and staring intently towards the tigers, accurately indicating their direction, most likely by using their highly developed sense of smell. Contrary to our wishes the Chinese built up the fire. The flames rose high into the dark sky, sending up showers of sparks.

Whether because of the hissing flames - which all animals fear - or the presence of people, the tigers were scared off; they did not attack the tethered and defenceless horses.

Standing by them, we heard branches cracking and leaves rustling under the heavy footsteps of the massive cats, but we failed to catch sight of them, although Dolgikhk claimed that he made out the shape of a tiger in a moonlit clearing.

The tigers' voices sometimes came closer, sometimes died away in the depths of the taiga. From the loudness and timbre of the calls it was easy to tell that there were two animals: one roared with a deep bass pitch, sounding like 'oon-oon', whereas the other's calls resembled an ordinary cat's growling and miaowing, only louder. They were clearly male and female.

This was confirmed the next day when we followed the spoor in the snow: the male's is more rounded, and the impressions of the toe pads in front are blunter than in the female; her spoor is somewhat elongated, with the toe pads narrower and sharper.

To search for the tigers in the thick undergrowth at night, even if moonlit, would have been pointless, since the cunning and wary animals keep well away from armed people. Hardly had we advanced towards them than they fell silent and went off into the undergrowth, making quite a noise in the process.

Before dawn, at about five o'clock, the tigers' voices went quiet and they evidently went away altogether. The horses became a bit calmer and began to eat straw. The sleepless night began to tell on us, and Dolgikh and I lay down on the platform beside the Chinese. We immediately fell fast asleep - not for long though, since in an hour at most we were woken by neighing, the clip-clop of hooves, and the shouting of the Chinese.

Suddenly the foreman, excited and panting, rushed into the cabin shouting: 'Captain, captain! Go up the mountain at once, a tiger is trying to attack the horses up there! Our people are scared!'

We needed no more encouragement. Seizing our rifles and ammunition, we ran outside. It was getting light. The Chinese were bustling about with the harnessed horses, shouting and gesticulating, and pointing to one whose neck bore bloody scratches. Apparently the tiger had attacked the horse train being taken into the forest to collect firewood. Emerging from behind a felled pine, it lept at a leading horse, but the shouts from the Chinese and the foreman's shots scared it so that it let go of its victim. It slowly made off up the slope, bad-temperedly growling, and took cover in a thicket. We had to run one and a half versts to reach the spot.

From the tracks we could see that the tiger had been lying in wait on the steep slope. When the horses came near it crawled from the pine over to the leading pair of horses and then rushed at the right-hand one: it did not need to leap, since the horses were right inside the ambush.

Deep down we were frustrated with the situation, since the tiger would have returned to the horse it had attacked and could then have been killed. Now, however, it might go far away, convinced that obtaining horse meat was too dangerous. Nevertheless we hurried off in pursuit of the tiger. Its track climbed a deep gorge towards the summit of the Blue Mountain. Thickets and boulders filling the narrow defile meant that we could not see more than fifty paces in front of us, so we moved quickly in the hope of overtaking the tiger before it reached its lair. Our strenuous climb and sleepless nights had exhausted us, and we had to stop frequently to rest.

At one spot, under a large old pine, we found a wild pig's lair lined with brushwood and dry leaves. The spoor was not fresh, and recent snow had already partly covered it. From this taiga bed I noticed, about ten *sazhen*[11] above us at the top of the slope, a patch of yellow. After a moment or two, I could clearly make out a tiger's head, neck, and part of its striped back. My rifle lay behind me. Without taking my eye off the tiger, I reached for it; but just at that moment my companion lept to his feet and cocked his rifle. That was enough. In a split second the animal disappeared from view.

By the time we had scrambled up the slope the trail had gone cold. It was clear that the tiger had been keeping us not only within earshot but also 'within its sights'. Further on it had proceeded further up the ridge in a series of jumps. We could hear it ahead of us, crashing through the vegetation. Not yet giving up, we continued to follow on its heels while carefully examining the forest ahead.

By noon we were about half way up the mountain. Above us the slope became much steeper, rocky, and in places precipitous. Overhanging cliffs soared up fifty *sazhen* or more. The tiger had climbed higher and was now up there somewhere. How could we get at it? After pondering the situation we decided to detour around the cliffs and climb above them from the other side, where the slope was not so steep and rocky.

Leaving the tiger track on our right we began to work around the granite cliffs in front of us. Here among scrub and rocks we found tiger paths in the snow. We were nearing the top of the ridge when we heard, from the other side of it, sounds of something crashing through vegetation. It was our tiger descending the far slope. Our frustration was unbearable! Again the elusive animal had made its escape, to wander once more in freedom through the boundless expanse of the taiga.

On the far slope of the ridge, where the tiger had made its escape, we found its lair under an overhanging rock. The lair was four paces long and two wide, and was lined with dry leaves. It was still warm. The tiger spoor led away down to a stream, then climbed again, on and up towards the summit of the Blue Mountain, until it was lost among boulders and thickets. Disappointed at our lack of success we set off back down.

Night had already fallen when we reached the cabin. There was a bright moon in the dark sky, and the stars were twinkling overhead. The cold blue peak of the mysterious and regal mountain shone high above us. The taiga was sleeping, quietly breathing. From afar came the rustle of dry foliage and the sound of the mountain air blowing through the canopy of the forest giants.

Chapter 11

LIFE CYCLE, PREDATION, COMPETITORS

Our account of Amur-tiger ecology begins with their reproduction. Tigers tend to be solitary, though males may sometimes feed or rest with females outside the mating season. Male tigers know when a female is ready to mate (is 'on heat', or comes into oestrus) by changes in the scent she uses to mark her territory. Females may show interest in the male's footprints and scent markings; she may also roar or give other calls so that the males in the area are soon attracted to her. A male tiger will mate with as many females as he can find on heat, but will normally reach an oestrus female first only if her territory is within his own.

Amur tigers mate in the coldest winter months - from December to February. Baykov described how 'In the cold, clear January nights tiger duets and trios may often be heard in the dense pine forests of Lao-e-lin, resounding clearly through the mountain air. The call of the tiger has a fearful effect in the great, wild, primordial forest. At night it inevitably arouses visions of the terrible might of the colossal cat'.[1]

If an oestrus female is not ready to mate she will be aggressive, so a hopeful male usually approaches cautiously and stops some distance away. Generally they are not strangers: their territories overlap and they regularly communicate through scents and calls. He moves closer by stages and lies down after each advance. Eventually the female co-operates and the two move closer together.

The procedure is tentative, and there may be many withdrawals before physical contact is made. Courtship play follows, which involves pawing, licking and cheek rubbing, as well as quiet sounds such as grunting and the tiger's typical greeting - a nasal puffing.[2] Play-fighting may develop, and in the Russian Far East areas of flattened snow have been found at mating sites.

Finally the tigress assumes a special crouching position, signaling her readiness. The male mounts her and pinches the scruff of her neck with his incisor teeth as they copulate.

Tigers mate frequently - even hundreds of times - over several days. During this time the female may become aggressive again and viciously swipe at the male, especially after each copulation, so that he has to be wary. Usually the pair go their own separate ways once the 7-14 day oestrus period has ended, though sometimes they stay together longer. Certainly tigers have been found to socialise at kills more frequently than was once believed.

The tigress will come on heat again after about five weeks if she has not conceived or if a litter is lost. If she is pregnant however the cubs will be born after a gestation period of 14-15 weeks.[3]

Family Life

The cubs are born blind and helpless, weighing only 1.5 kg. The cubs' eyes open after one or two weeks; initially blue or blue-green, they change later to glowing amber. Usually there are two or three cubs in a litter, though single cubs are apparently common among Amur tigers. The reproductive rate of Amur tigers seems to be lower than that of other tigers, probably because of the lower prey density.

Tigresses are devoted mothers. The cubs are kept safely in the den, usually among the roots of a fallen tree or in a cave among rocks. She will move them to a place of greater safety if there is any disturbance, carrying them gently one by one in her huge jaws. Perhaps as many as half the cubs perish before their first year is out. Brown bears and especially other tigers are a potential threat. The tigress must choose a carefully hidden den and leave the cubs alone for as short a time as possible while she hunts. Forest fires and grass fires - often started deliberately to improve grazing in the more settled areas of Primore - are a danger to cubs.

The cubs remain in the den for 4-8 weeks, and then will venture out for short distances. They make their first excursions away from the den accompanied by their mother when they are 3-4 months old, keeping in single file behind the tigress, and it has been suggested that her striped tail and large white spots behind the ears act as beacons for the cubs to follow. Soon after, they are regularly at her kills, but still hide while the tigress is hunting.

It is extremely rare for more than two cubs in a litter to survive to maturity, though the survival rate may be higher if prey is abundant. The ratio of male to female cubs born is about equal. Males leave the family earlier than females and are more likely to be killed by older territorial males, or to perish because of their inexperience in hunting. Being less cautious they are more likely than more experienced adult tigers to become victims of poachers. So generally females survive longer than males.

Tiger cubs are playful. Their games improve their strength and co-ordination and teach them the skills necessary for survival. They stalk and pounce on leaves, insects and even their mother's tail. They are soon wrestling with their siblings, and even with their mother as they grow older. Sometimes however play among older cubs may be too rough: in 2002 a 24-month-old cub was found dead in Lazovski Zapovednik - accidentally killed by a neck bite from one of its litter mates.

At first the cubs must hide in the undergrowth while their mother hunts - as they may give her away to the prey being stalked. Later they are allowed to watch, and eventually join in. After one year of age cubs accompany the tigress nearly all the time. They may help her by driving the chosen victim towards her.

Learning to hunt is a difficult and dangerous process. Cubs may be killed by the flailing hooves of *izyubr* or by the tusks of wild pigs. Inexperienced cubs tend to grab the legs of the prey, leaving them open to retaliation. Sometimes the mother will intervene by bringing the prey down and then leaving it for the cubs to kill. It takes many attempts before the cubs learn to kill efficiently.

A mother tiger may allow the cubs to feed first. If she joins her young at a meal she will withdraw if a cub protests and will go without meat to ensure that they have enough. The cubs have voracious appetites, and by the time they are a year and a half old the males especially are about as large as their mothers and it is a strain for the tigress to capture enough prey. The weakest is last to feed and if there is insufficient meat it will starve to death. Male tigers do not generally help rear the young and usually have no contact with them.

By 18 months or so, when they have their permanent canine teeth, the cubs are starting to move and hunt independently of their mother. Male Amur tigers usually leave their mothers at about two years and females up to a year later. A female is usually ready to produce cubs when she is 3-4 years old. Sub-adult male tigers are sometimes tolerated by older males, but this will change when they reach maturity. Young males may look for a vacant territory or one where there is a chance of ousting an old or sick male. The top cat does not usually retain its dominant position for many years before being challenged by a younger and stronger rival, but during that time the male's dominance gives stability to the tiger families within his territory. Other male tigers taking over the territory may kill cubs sired by the previous incumbent in order to bring the female into season. Ousted male tigers may be killed or injured. A wild tiger is usually past its prime by 12-13 years, but may live to 15-16 years.

Territories and marking

Like the other wild cats of the region - leopard, lynx and forest cat - tigers are essentially solitary hunters with well demarcated territories.[4] Amur tigers need large territories, and that of a dominant male usually overlaps those of several breeding females and their progeny; the larger his territory, the more females are accessible to him.

The sizes of the territories vary with the density of their population. In Lazovski Zapovednik, where the tigers are more numerous than in most parts of their range, each female's home range is about 100 sq. km. In other areas females have home ranges of around 400 sq.km and resident males of 1,000 sq.km. In the zapovednik one resident male's territory includes those of four resident breeding females. The need for a great deal of space and good-quality habitat inevitably leads to competition with human needs, so that problems are likely to arise.

Flehmen response
The grimace, often made by a tiger sniffing its own or another tiger's urine, involves half closing the eyes, drawing back the lips, wrinkling the face, and hanging out the tongue.

Marking is an important factor in the spacing of the tiger population. Scent marking forms the basis for territorial behaviour as it enables tigers to keep track of each other's movements and avoid conflict. To acquire information from scent the tiger usually hangs out its tongue and draws back its lips in a grimace, causing its eyes to close. This is called the flehmen response. It allows the tiger to pass the scent through two small holes in the upper palate behind the incisor teeth: in effect, the tiger can taste the scent. It looks like an expression of disgust!

Using trees and rocks tigers mark their territories and routes with small quantities of urine - as do domestic cats. These contain pheromones and hormones - and also amine, which gives the secretion its unique scent. Some of these chemicals are special to each tiger and indicate, for example, how recently a particular tiger passed by, whether it was male or female, and whether the female was on heat. Tigers mark most frequently when they are establishing territories, especially near territorial boundaries and contact points such as crossings. A female marks most frequently just before coming on heat, and the male also marks frequently while she is on heat. The strong scent may persist for a long time and is regularly renewed.

Tigers rub their cheeks - where there are other glands - on trees and other objects that they have already sprayed with urine. In addition trees are marked by the tiger raking its claws along the bark, while almost hugging the tree, leaving scent from glands on their paws. They do this usually on trees near a kill.

Distinctive scrapes are left on the soil or snow. These seem to have nothing to do with concealment of excrement, since faeces and urine are left on the surface - the former in prominent piles that may also convey a message. Males seem to 'scrape' more than females. Tigers also roll on the ground in order to leave their body odour to mark their territories.

Intervals between marking are fairly regular, but more frequent near kills and near the tiger's lair. Marking conveys messages to other tigers, especially at shared crossing points, indicating the presence of the resident tiger. It seems to have a general communication purpose, not just to demonstrate territory. Apart from when the female is on heat, the marking and roaring - something like a long drawn out 'aa-oon' - seem to be saying 'I am here; keep away!' Maintaining territories requires constant activity for both males and females, and territorial boundaries tend to change.

Trails and lairs

From tracking tigers in the snow and mapping their routes, Russian field workers have discovered an undoubted regularity in their movements. Tigers are creatures of habit. When not hunting or seeking shelter, the animals will often move for up to 10 km at a stretch in a more or less straight line - usually following a river valley, the bed of a frozen stream, the foot of a slope, a ridge or watershed, or a trail made by other animals or even by humans. Tigers prefer to follow routes on south-facing sunny slopes, especially near the crest of a ridge, so avoiding the steepest slopes and places where the snow is deepest.

At other times tigers may retrace their tracks as a defensive and precautionary measure if they sense danger, when they leave their dens or in hunting manoeuvres. Double tracks and even quite complex loops are often observed.

For a long rest during the day the tiger usually selects somewhere giving a good view of the surroundings. Wild pigs' nests, hollows in the snow, rocky recesses and niches, and hollows under fallen trees are often used. The tiger regularly visits these beds - important staging posts on its routes. When close to their kills tigers may lie up nearby, near bushes or tree trunks for protection. They enjoy basking in the sun and may lie out in quite open places. Their real lairs are well protected resting places that they approach with great care, their tracks often showing a sharp change in direction or some other complicated manoeuvre.

Several lairs have been described by Evgeni Matyushkin in Sikhote-Alin Zapovednik.[5] One was on a stone ledge, covered with dry pine needles and soil, near the foot of a steep cliff. This cliff was part of a jumble of rocks on the ridge of a spur, approachable only from the side. The lair was about 150 m above the bank of the river, with a commanding view. The ledge faced southwest, well warmed by the sun and protected from northerly winds. Snow was unlikely to penetrate and the overhanging rock kept off the rain. The large male occupant of this desirable residence visited it twice during two winters, each time approaching by a different route.

A stable network of tiger trails is useful as the tiger can reach its destination with minimal expenditure of energy - especially important in winter. It can move quickly to favourable hunting areas, or to meet potential mates. The trails tend to keep to a constant level of altitude and depth of snow cover. Tigers travel during both day and night, often covering 15-20 km in 24 hours, or even much further. They travel in a measured way, as they like a lot of rest - usually lying down 6-8 times in every 10km.

Hunting

Tigers are the top predators in the forests of the Russian Far East. Like other predators they will catch any animal they can, and will pursue whatever prey is most numerous in a given area. On the grounds of energy efficiency tigers do show a preference for the largest prey animals.[6] These are generally *izyubr*, wild pigs, and sika deer, and also roe deer, gorals, elk and musk deer. Amur tigers, especially cubs, also kill smaller mammals such as Manchurian hares, raccoon dogs, badgers, rats and mice, as well as birds such as pheasants and hazel grouse. With a shorter growing season than in southern Asia, the tigers' prey in the Russian Far East is more thinly distributed and so tigers need much larger territories for their survival than, for example, in the tropical deciduous forests of India.[7] Before human hunting of tigers became significant in the 19th century the range of the Amur tiger was largely determined by abundance of prey.

The great striped cats depend on their excellent senses of smell, hearing and sight to detect prey. They have keen eyesight and, since the eyes face forward - giving them stereoscopic vision - they can judge direction and distance very accurately. Night vision is excellent, and hearing is especially acute: their ears are constantly attuned to the variety of sounds and can distinguish between an animal's footfall and a naturally falling leaf. Amur tigers hunt mostly in the evening, night and early morning.

A prey animal may be smelled or heard before it is seen. The potential victim is then approached and carefully stalked, following a zigzag course. When the hunter is as close as possible, there comes the charge - a sudden quickening of pace followed by a number of bounds. As a rule the tiger does not attack its intended prey from a distance greater than 30 m, and more often it is from 10-15 m. The chase will not usually be continued after 150-200 m. If the hunt fails the tiger does not always give up completely: it may lie down and rest before repeating the attack at intervals of 1 km or less.[8]

As tigers attack from the rear, they usually grab the prey in the hindquarters with their powerful claws and are sometimes pulled along by it. This leaves a clear track in the snow. The tiger pulls the victim off its feet, avoiding any flailing hooves, then moves up its body and closes with a killing bite to the back of the neck, or to the throat after pulling its head sideways. The tiger maintains a suffocating grip on the throat, in many cases for several minutes, until the prey stops struggling.

Amur tigers are immensely powerful. They can kill adult wild pig, *izyubr* stags, elks, and even bears. Tigers walk on the fore pads of their paws, which makes their stride supple and elastic. These powerful flexible fore paws and forelimbs can be twisted inward, allowing them to grasp their prey. Their razor sharp claws remain retracted until they are needed in the final moments of the attack.

Tigers have huge canine teeth which deliver a fatal bite to the victim's throat or the back of the neck at the base of the skull, wedging apart and severing the cervical vertebrae. Alternatively the windpipe is crushed and the main arteries are severed. Death usually follows very quickly, or else the victim may die of suffocation.

Prey is seldom eaten where it is killed: it is dragged away to a more secluded spot - under a fallen tree or in the lair of another animal such as a wild pig. Tigers can drag prey weighing 250-300 kg for a considerable distance: the carcass may be moved as far as 40 m initially and then several more times for later meals.

Tigers have sharp carnassial teeth that act like scissors to cut and slice through skin and tendons. The first parts to be eaten are the hindquarters and part of the intestines. Several hours or 5-7 days may be spent near the kill, and a tiger can consume a great deal of meat in one sitting. Yudakov described how an adult male ate every last scrap of a roe deer, at one sitting, without changing its position. However, tigers hardly ever stay near their prey until it is completely consumed: 20-70% of it may be left.[9] Although the tiger may return to the kill the remains are usually eaten by scavengers, or even by other tigers.

According to Yudakov (1973), over a period of 12 days one solitary tigress killed four wild pigs, one *izyubr* and one roe deer. In 40 days a male killed three *izyubr* and four wild pigs. From this information he estimated that an adult tiger kills at least 70-75 individual prey animals each year, whereas Kaplanov (1948) estimated that an adult tiger kills annually the equivalent of about thirty large prey animals, each weighing up to 100 kg. Yudakov's estimate includes younger or smaller animals, which may account for the apparent difference.

Within reserves the deer and wild pig populations of the Russian Far East have been stable or have increased slightly in the late 1990s. However, outside protected areas the tigers' prey is thought to have seriously declined because of over-hunting by humans. This trend may be critical for the future of the great cat: it is unlikely to maintain its present population with the current numbers of deer and wild pigs. The current number of ungulates in the Russian Far East must increase at least threefold to support 450-500 tigers.[10]

The range of the Amur tiger in Khabarovsk Region is clearly limited to that of *izyubr* and wild pigs. In the south, the relatively high density of tigers in Lazovski Zapovednik is linked to the large numbers of sika and roe deer that supplement the *izyubr* and wild pig in the tiger's diet. The tigers in the reserve depend heavily on wild sika deer, but otherwise the seasonal diet is similar to that in Sikhote-Alin Zapovednik. Wild pigs, which do not cope well with deep snow,[11] are more commonly taken in autumn and winter when they gather where there are plenty of fallen pine kernels and acorns. *Izyubr* - more mobile in snow - are pursued more often in the spring and summer, especially at salt licks.

Tigers will sometimes kill smaller mammals and birds as well as fish and turtles. Baykov observed tigers fishing in streams - rather as bears do. In the winter of 2002-03 a year-old orphan successfully hunted young harbour seals along the coast of Sikhote-Alin Zapovednik: he killed more than seven of them, started to eat them from the head end, and then neatly stacked the bodies for future use close to the cliff face, away from scavengers.[12]

Tigers are not always successful hunters, and sometimes eat carrion, or even vegetable matter - including pine kernels and fruit. In one instance two starving tigers were found to be painfully thin: they had nothing in their stomachs except lichens.[13]

Prey

In many parts of the Amur tiger's range its most important prey species is the **wild pig**[14]. Adult pigs are dark brown with a whitish moustache, while piglets are striped. Weighing up to a quarter of a tonne it is the largest of the wild pig subspecies. Its distribution in the area is limited to the Manchurian-type forest, and also by the depth of snow cover.

Tigers are the main natural enemy of wild pigs in the region. They may approach a sleeping or foraging group of wild pigs at night to within 10-12 m, and then launch an attack on one of them. In one incident wild pigs moved just 200-300 m from the site of the attack, and stopped for the night again. In another case a pigs' family bed was found only 400 m from where a tigress was feeding on an adult wild pig over a two-day period.[15] As the tigers tend to follow groups of wild pigs, picking off individuals, they were known as 'herders of wild pig'.

However, wild pigs are difficult to tackle: they can defend themselves fiercely with their sharp tusks and also move with surprising speed and agility. In one case Bromley came across the corpse of an old pig that had been killed by a tiger. Later he found the body of the tiger some kilometres distant. It had died of wounds sustained in the Battle Royal.[16] In a more recent incident in Ussuriyski Zapovednik a tiger and a wild pig were both injured, after which the remainder of the herd turned on the tiger, killed it and ate it.

Although wild pig populations are subject to wide fluctuations and there can be heavy mortality in some years because of diseases such as swine fever, they have strong recuperative powers. However, the situation deteriorated sharply in the 1980s and '90s: the main threat has been over-hunting for meat, but widespread felling of Mongolian oak and Korean pine has also reduced their food supply.[17]

In the severe winters of 1982-84 and 1988, heavy snowfall limited access to acorns and other forage, and widespread disease reduced pig numbers. In such conditions only a few sows will produce litters, of perhaps only 2-3 piglets, whereas in favourable seasons with heavy acorn crops 90% of sows will have litters - often of 6-8 or even 10 young. In 1988 the depth of the snow cover in the Bikin valley was 90-120 cm and during that winter about half of the wild pigs were killed by tigers.[18]

The Korean pine is a key species in the mixed forests of Primore and Khabarovsk Region since wild pigs rely on pine kernels, especially in winter. They also feed on acorns. Both pine and oak show considerable variation in crop production - on account of weather, diseases and pests - and so large areas of mature forest of both pine and oak are necessary to reduce the impact of crop failures. Also eaten are nuts, berries, roots, fruit, horsetail rhizomes, fungi, worms, insects, carrion, eggs, fledglings, small mammals and even fish washed up on the shore.

It is clear that as long as so many local people are unemployed and do not have a reasonable standard of living they will continue to gather pine

Two izyubr *at a salt lick. The wet mud is salty, and provides the deer with essential nutrients.*

nuts from the forest by the tonne, even camping out in the snow in winter to collect them. They then sell them to Chinese dealers at such centres as Ussuriysk. The result is that the pigs' food source is further depleted. When many pigs left Sikhote-Alin Zapovednik during the 1970s, during a scarcity of pine nuts, the tigers did not seem to have followed them; instead they preferred to hunt *izyubr*, even in places where some wild pigs were still present. As the numbers of both these prey species have declined, mostly because of human activity, the prospects for their main predator, the Amur tiger, have worsened.

The ***izyubr***[19] - a close relative of the European red deer - is second only to the elk as the largest of the local deer, and like the wild pig can reach a quarter of a tonne in weight. Of a russet-red hue in summer its coat in winter shades to grey-brown. The reddish-orange rump patch is a prominent feature.

In the 19th century the *izyubr* was abundant and of great importance to the aboriginal tribes of the region, as their flesh, hides, tails, penises and especially the antlers 'in velvet' - used in traditional Chinese medicine - were paid as tribute to the Chinese imperial authorities and later traded with Chinese and Russian merchants. Although some were reared on deer farms for their antlers, in the early years of the 20th century great numbers of wild *izyubr* were slaughtered and the population fell sharply.

The main threats to the *izyubr* have continued to be excessive hunting and loss of habitat. With the economic crisis following the collapse of the Soviet Union in 1991 there has been a great wave of poaching with a corresponding drastic reduction in their population.[20]

A recent study has shown that the Amur tiger's distribution is more closely associated with that of the *izyubr* than with any other prey species.[21] Although the tiger is found almost everywhere that wild pigs survive in the Russian Far East, their distribution is patchy whereas the *izyubr* has a considerably greater range within the region. So the future of the Amur tiger depends to a considerable extent on a healthy *izyubr* population.

In Lazovski Zapovednik *izyubr* numbers fluctuated from 300 to 500 in the 1980s and the remains of 20-25 of them killed by tigers have been found each year. Tigers will frequent burnt areas, salt licks and the forests in river valleys where the deer are numerous. When the deer move down to the sea to drink salt water and feed on seaweed for their minerals, they are sometimes cornered and attacked by tigers in the raging surf.

Izyubr habitat differs from that of wild pigs, and felling of Korean pine probably has less effect on the deer. Indeed the removal of the canopy may stimulate the growth of the under-storey, and the deer may flourish where young forest growth provides forage, shelter and cover. Damaged forests, then, actually attract deer, but at the same time the forestry roads provide access for hunters.

Sika

This hind has no antlers, but the antlers of the stags are similar to those of izyubr. *In winter the distinctive spotted coat is replaced by a unform brown colour.*

One of the most attractive deer in the Russian Far East is the **sika**.[22] In Russian it is the 'spotted deer'. The Chinese call it *mei hua lu* - the flower petal deer. It has a bright chestnut-red summer coat with clear white spots and a light rump patch. In winter its coat is a dappled greyish-brown. They have simpler and less massive antlers than *izyubr*. The Russian Far East population is the largest and most heavily spotted of the sika deer, but it still reaches a maximum of only 140 kg in weight. Although hardy, sika deer are less tolerant of deep snow than *izyubr* and penetrate no further north than Sikhote-Alin Zapovednik, where there are some 120. They prefer secondary forest and burned areas with new growth of oak, hazel, maple, lime, aralia, and herbage that provide rich feeding.

Sika deer have been persecuted by human hunters even more mercilessly than have *izyubr*, especially for antlers in velvet - *panty*, valued in traditional Chinese medicine as a tonic - as well as for their meat and hides. By the 1940s they were very rare with fewer than 300 surviving in the wild in Primore, though a much larger number were reared on deer farms and formed a valuable part of the regional economy.

With strict protection the number of wild sika in Primore increased.[23] With 800-1000 in Lazovski Zapovednik alone, this has resulted in damage to coastal vegetation - especially at one point where the deer congregate in summer to avoid blood-sucking insects and to drink salt water. These gatherings attract tigers, which may take advantage of sea fogs to stalk their prey. Although sika are found only in a small part of the Amur tiger's range, they have increased in Lazovski, Ussuriyski and Kedrovaya Pad zapovedniks, and are now the tigers' most important prey in these areas.

Siberian roe deer

The Siberian roe is quite similar to the European species - small, with a barely visible tail, and antlers which are woolly during winter while growing, to be ready for the early rut.

The light ochre-grey winter coat of **roe deer**[24] gives way to a bright russet-red in summer. It is one of the most numerous deer in Russia and China, and weighs up to 40 kg. It feeds on fresh grass, shoots and buds in spring, growing leaves and grass in summer, and acorns, nuts and fruit in autumn. Roe are hunted extensively for meat and hides, and their numbers have accordingly been dramatically reduced. Lynxes are their most important natural predators in the Russian Far East although Amur tigers and leopards also take many, while wolverines and yellow-throated martens may prey on fawns. In winter, especially, many succumb to wolves and feral dogs.

Amur tigers also prey on **musk deer**.[25] It is a very small deer-like animal, rarely reaching 17 kg, and of a chocolate-brown tawny colour with grey spots on the upper parts.

Musk deer
This curious animal is not strictly a deer at all, but belongs to a family of its own (Moschidae). It has no antlers, but instead the male has two tusks in its upper jaw. The hind legs and rear half of the body are much larger than the front half. Glands on the chest produce the valuable musk scent.

It is remarkable for not having antlers: instead both sexes have enlarged canine teeth, which are more prominent in the males. It has a rather ungainly appearance, with a small head, short neck, large ears and long limbs - especially the hind ones. These give it a singular kangaroo-like gait and enable it to reach for lichens festooned on branches. It also feeds on mosses, herbage and the shoots of fir, spruce and other trees. Favouring the dark Okhotsk fir-spruce forests, and so not always found in the tiger's optimum habitat, it does not form a large proportion of the tiger's diet.[26]

Gorals[27] - relatives of the chamois - are remarkable denizens of the cliffs along the Sea of Japan and are occasional prey for Amur tigers. Looking like sturdy goats, they are light grey, sometimes brown, with a black stripe along the back; the beard and long tail are white. Both sexes have short curved horns - larger in the males - which can be used to good effect against predators.

Favoured habitats are the thickets of stunted oak, birch and pine at the summits of the sheer cliffs on the coast. The gorals emerge onto the cliff tops at daybreak and have been given the name of 'dawn watcher'. There they feed on the scrub Mongolian oak, hazel, maple and birch, as well as herbaceous plants. The Amur goral is now a great rarity.[28] Along the coastal cliffs in Sikhote-Alin Zapovednik lynxes are the main natural predators of gorals. They are also taken by white tailed and Steller's sea eagles, wolves and sometimes by tigers.

Evocative of the northern taiga is the **elk**[29] - the largest of the deer family, though in the Russian Far East the elk are smaller than other races of the species, weighing up to 400 kg. Its coat is almost black in summer and rusty brown in winter. Although ungainly in appearance with a hump-shaped muzzle, long ears, long legs and a short heavy body, the elk is very fleet of foot. It is found in the northern part of the Sikhote-Alin, as well as in the north of China's Heilongjiang Province and in northern Mongolia.

Asiatic black bear

This species is the commoner of the two bears in the mexed forests of pine and deciduous trees. It is easily distinguished by the black fur and the white or yellowish V-shape on its chest. Its ears are also very prominent. It readily climbs, and often makes its home inside a hollow tree.

Elk prefer river valleys and lowland marshes in the coniferous taiga belt. They usually avoid mountainous and open areas, although they may sometimes move to higher altitudes to escape irritating flies and midges in summer.

In Sikhote-Alin Zapovednik, where there were only 30-35 elk in the 1990s, they are found especially in the northwestern river valleys in the coniferous taiga and marshes - which is not optimum tiger habitat. Tigers regularly prey on elk in the northern part of the Sikhote-Alin but the great deer are found only in a small part of the Amur tiger's range.

The Asiatic **black bear**[30] is smaller and much less powerful than the brown bear. Its head is more flattened and narrow, with a more pointed muzzle and larger ears. The bear takes to trees more readily, but eats less meat. Its black coat is thick and glossy, with a pale chevron on the chest. In the Sikhote-Alin the black bears inhabit mixed broadleaf-conifer forests - especially in river valleys where there are stands of Korean pine.

Apart from humans, the Amur tiger is its main enemy throughout its range in the Manchurian-type forests. Hunters in the Russian Far East frequently find the remains of black bears killed and eaten by tigers, and they often occur in tiger faeces. According to Baykov, tigers can handle bears, black or brown, of almost their own weight, stalking them or ambushing them from a cliff or fallen tree. When an unsuspecting bear ambles into such an ambush the tiger springs onto its back from above, seizing it with one immense paw under the chin and the other around the neck, biting through the neck vertebrae. Should the black bear escape it will try to take refuge in a nearby tree, in which case the tiger may feign a departure and hide up nearby until the bear descends; then it renews the attack. When the bear tries to escape by climbing it is sometimes pulled down by the tiger. The tracks and bed of a tiger were once found under a poplar tree, with a bear hiding in its crown - indicating a protracted siege. Predation by tigers in Lazovski Zapovednik however did not seem to reduce the population of the bears in the 1970s.[31]

Brown bears are also taken. Though tigers will usually avoid the formidable large males, they regularly prey on cubs, and occasionally adults. In Sikhote-Alin Zapovednik a large male tiger made this his speciality. Brown bears are sometimes taken from their dens in winter by tigers.

Natural Competitors

The success of a carnivore may be affected by the presence of natural competitors. The other leading carnivores of the Manchurian region - the Far Eastern leopard, brown bear, wolf and lynx - do compete with the Amur tiger for prey to a limited extent. But these hunters are all part of the rich ecosystem of the Ussuri taiga and face similar threats from human activities. The conservation issues affecting the Far Eastern leopard are very much the same as for the tiger.

The Far Eastern or **Amur Leopard**[32] is a strikingly beautiful cat with a luxuriant winter pelt - typically a handsome pale straw yellow - and a long thick tail. In summer the shorter fur sometimes has a reddish tinge.[33] The spotted pattern consists of large black rosettes that may form solid rings. Although the leopard is only a tenth of the Amur tiger's weight, it is a powerful predator with a lithe supple body.

Rather overshadowed by its larger cousin it is even more endangered and shares but a small part of the Amur tiger's range. In 2003 around 30-40 Amur leopards survived compared with some 400 Amur tigers. The area inhabited by the leopard in the Russian Far East is only about 2600 sq.km compared with around 150,000 sq.km for the tiger - like comparing Dorset with the whole of England.

The very few surviving wild leopards prowl the forested mountains where the frontiers of the Russian Far East, Northeast China and North Korea meet near the Sea of Japan. In Primore the leopard is at the northern limit of its range: its former strongholds were originally in Northeast China and Korea. Its preferred habitat is forests of Korean

Leopard

Rather smaller than the tiger, the Amur leopard is much rarer - on the verge of extinction in the wild. It readily climbs trees, and a leopard high on a branch can sometimes be spotted only by the twitching of its tail.

pine, black fir and broadleaf trees where there is sufficient prey - roe and sika deer, Manchurian hares, badgers, raccoon dogs and pheasants.[34]

Leopards avoid tigers, and the two great cats can co-exist partly because leopards climb trees easily and partly because the tigers tend to prey on larger animals. Leopard territories are smaller than those of tigers, though they may overlap. Both animals are essentially solitary except during the mating season and - as in the tiger - an adult male leopard's territory may include those of one or more breeding females.

The Udege and Nanay - aboriginal tribes of the region - regarded the leopard as *jarga*, a totem animal. Following the increase in Chinese and Russian settlement, especially with the building of the Trans-Siberian Railway in the 1890s, leopards were killed in greater numbers for their precious skins and body parts, especially the bones, used in traditional medicine. Many were shot or snared in retaliation for their depredations on deer farms, where the herds were - and still are - reared for their antlers. Even at the beginning of the 20th century Amur leopards were rare and there were probably only about 400 surviving, including around 100 on Russian territory.[35]

By the 1970s the range of the leopard in the Russian Far East had contracted and fragmented into three parts.[36] It has been suggested that this was caused by increased numbers of tigers in Primore since the 1970s, but it is more likely the result of human activities, especially poaching. Not only had their range in the Russian Far East halved from 1972 to around 2600 sq.km by 2000-02, there are even fewer leopards remaining in Northeast China - effectively part of the neighbouring Russian population. Deforestation, the demand for wildlife products for food and traditional medicine, and the widespread use of snares and poisons have had a devastating effect on the leopard and tiger populations in Northeast China, where they were both formerly widespread.[37]

The Amur leopard has been protected in Russia since 1956, and capture for zoos was restricted in 1966. However, that has not prevented a decline, mostly as a result of poaching. This continues: six leopard skins were confiscated in 2002 in southwest Primore. There has been a drastic reduction in the numbers of its prey - notably roe and sika deer - by human hunters, compelling the leopards to attack domestic stock, especially farmed deer. Large areas of leopard habitat have been destroyed as forests in the southwest of Primore have been logged, burnt or cleared for pasture.[38] In addition there has been disruption of wild places as a result of mineral extraction and extensive road building.

Leopard conservation measures include anti-poaching patrols and customs control, fighting forest fires, fostering roe and sika deer populations through additional feeding in severe winters, and paying compensation to farmers who have lost stock to leopard and tiger depredation. One option being considered is the reintroduction of captive Amur leopards into the wild, but to be successful there has to be suitable habitat and sufficient prey. In the meantime the tiny population of the last wild Amur leopards is being closely monitored.

Manchurian **brown bears**[39] are the largest carnivores in the Russian Far East. These tremendously powerful animals may exceed 400 kg in weight and in the Sikhote-Alin they are often dark or almost black in colour. Brown bears are significant predators of *izyubr*, and will also feed on deer killed by tigers. They even occasionally kill and eat young tigers, and very rarely an adult one. They also consume tigers' leftovers and sometimes deliberately follow tigers for this purpose. However, being omnivorous they do not always compete with tigers for food. So brown bears may be victims, competitors and even predators of the Amur tiger. In general the two species seem to maintain a degree of cautious rivalry and wary co-existence.

Wolves are another significant predator of *izyubr*. In the early 20th century, when tigers were more numerous, wolves were not frequently encountered in Primore. As tigers and leopards became rarer wolves spread through the region, but they kept in the main to forest around populated areas where animal husbandry is practised. Only in winters with little snow did they migrate into remote forest areas. In Russia wolves have been much persecuted by sport hunting and extermination programmes. In the 1970s there were about 250 in Primore and around 300 in Khabarovsk Region, but since then their numbers have been reduced. In Sikhote-Alin Zapovednik wolves prey mostly on *izyubr*, the main prey of the tiger. However, they give way to the tigers and usually leave areas where tigers are resident. In any case they tend to occupy more open country and the dark coniferous forests where tigers rarely go. Wolves are indeed occasionally killed by tigers, although there have been few recorded cases.[40]

Tigers mostly depend on large hoofed animals for prey, so lynxes, Far Eastern forest cats and yellow-throated martens compete with them for prey only to a limited extent. It is true that lynxes sometimes prey on roe deer, gorals and musk deer, but rarely on adult sika or izyubr, or wild pigs. The Far-East forest cat mostly takes hares. Yellow-throated martens hunt musk deer and - like sable and Siberian weasel - also prey on squirrels and other rodents. Increased protection for the tiger in the Russian Far East may also afford a respite for all these impressive carnivores.

Chapter 12

CAPTURING TIGERS[1]

(Note: Tigers used to be captured by hand in the Sikhote-Alin Mountains for zoos. A number of families became famous locally for their exploits, including the brothers Kozin, Bogachëv, Glushak, Trofimov, Cherepanov and Kalugin, and other individuals such as Kondrat Shevkunov.

Ignat and Makar Trofimov feature in this story, although part way through Makar goes back to the village to take in the harvest, returning only at the end as they set off on a new hunt. The other expert catchers are Petrushenko, Kalugin and the author, and the rather inept Zabaluyev is the party's dog's-body. The actual dogs are Kado, Dik and Sharik.

The story describes the hunt for a tigress lacking a toe in one foot - Toeless - and her two full-size cubs. There has been a long, arduous and fruitless trek through the forests and mountains up to the point where we join the expedition. For the first day or two, only Zabaluyev and the author are in camp.)

'New Year's Eve'

The snow fell increasingly fast. The signs were clear: a blizzard was coming. Zabaluyev and I fitted up a shelter for the horse. Fortunately we had with us a horse cloth; that gave it some protection.

We had only just managed to drag up a dead elm tree for firewood when the first gust of wind burst upon us with a howl. The taiga groaned, as if some unbearable weight had been loaded onto its shoulders. Everything was whirling around us in a furious dance. Tremendous blasts of wind followed on each other's heels. Trees swayed back and forth, creaking, while the flying snow formed a dense wall. Now and then could be heard the crash of a forest giant falling not far away. It seemed as though the taiga was intent on burying us in its shifting snow drifts.

We sat by the red-hot stove with all our outdoor clothes on. The elements raged, playing havoc with our imaginations. Fighting off the fear, I recovered a sense of realism, and comforted myself with the thought that Ignat was a real expert at survival in the taiga.

'In his hard life as a hunter, he has lived through more than one blizzard,' I thought to myself.

The storm continued all that day and the next - which was December the 30th. Occasionally we went out to saw up a dead ash tree nearby. Then we'd brew up strong hot tea. We brought the dogs into the tent, so that they wouldn't be frozen.

During the night of the 30th there was no improvement in the weather. Only towards morning did the gusts die down a bit. At dawn it finally stopped snowing. When it was quite light, we crawled out of the tent. Enormous snow drifts were all around us. The dogs shook themselves and, to quench their thirst, gulped down dry snow.

I plodded over to the stream to fetch some ice. After twenty paces I came across some half-obscured bear spoor. The *shatun*[2] had evidently crossed the stream before dawn and, noticing the smell of the tent, had gone around it and then climbed the hill.

Back in the tent I told Zabaluyev about it. After breakfast he talked me into going after the bear. He said he had never yet seen a bear in the taiga. The bear does not generally run, so I knew it could not have gone far. It was a tempting idea, despite Ignat's prohibition of such excursions, and I agreed.

It did not take us long to get ready. In twenty minutes or so, with rifles over our shoulders and the dogs on leads, we were already climbing the steep slope through dense birch and aspen. The *shatun* had ploughed a deep furrow through the snow.

The bear began to take evasive action. I tried moving silently, so as not to alarm it. But it was hardly possible to stalk it in such a thicket! Zabaluyev began to flag and lagged behind.

Not quite at the top, in a tangle of vines, the bear got our scent. It made a noise like a steam engine and rushed headlong downhill towards the Khanikheza River. We released the dogs, which had been straining at their leashes. On the slope Dik started barking and, like a bell, Kado struck up too. I rushed down towards them. Nearby, Zabaluyev was running clumsily, breaking down bushes in his haste. It took me about three minutes to reach the scene of action.

The bear was in a gully, roaring and driving off the dogs clinging to it. Taking aim, I fired. With a roar it raised a cloud of snow and with one leap disappeared into the undergrowth. The sharp-pointed bullet had hit an aspen trunk, and ricocheted away harmlessly.

The dogs started barking once more at the river. I ran from the steep bank and saw what seemed to be a fresh *izyubr* track. But behind a mound of snow I saw tiger spoor! Just then Kado jumped out from undergrowth on that side. I called him, but he paid no attention, disappearing into the bushes. Suddenly there was a short deep roar. A dog's yelp reached our ears and over by a bend in the river the willows jerked and then swayed gently.

With racing heart I retraced my steps. Kado's fate was all too clear. What if Dik was there as well? Only now did I realise how wrong I had been to embark on this folly.

'Scaring away the tigers, *and* losing my best dog! How stupid!'

At this moment I saw Zabaluyev coming along the path. I explained what had happened and advised him to go back to the tent. I still nurtured the hope that the easily aroused Dik was occupied with the bear, and had not noticed the tiger spoor. I wanted to get away from this fateful place, and continue pursuing the bear.

However, Zabaluyev was afraid to go back alone. From now on he stuck to me like a limpet. Clutching his gun, he hardly dared move. After an hour and a half we found the bear's track. Climbing the northern slope, covered in dense pine, we reached the peak of the conical hill. Although we listened intently for a long time, the snow-covered taiga remained silent. Evening was approaching. We lit a fire and waited.

This last night of the year was very cold. The bright moon and untwinkling stars shone down from on high. Limpid light penetrated the depths of the forest, and the trees' shadows were cast on the pale shroud of snow. Pine branches flared up on the fire. Zabaluyev talked longingly about the festivities in town, with decorated fir trees and tables loaded with all kinds of food.

I, on the other hand, thought only about how Ignat would react to my misdemeanour, and how far away the toeless tigress and her grown-up cubs now were.

After about two hours, with tail down, Dik came to the fire. The poor dog looked exhausted. We gave him our last few rusks and set off for the tent. After giving the horse its oats, we turned in, tired out from our day's exertions and our failure. But it was not possible to sleep. Then a rustle could be heard, Dik barked, and when I looked out I saw Sharik running up to the tent. Kalugin's voice came to us. We lit a lamp.

By the tent, shaking snow off their sleeves, taking off their heavy packs and stamping their feet, were Ignat, Petrushenko and Kalugin. Dik barked at Volchok[3] and Pirat. When we had all calmed down, I plucked up courage and told Ignat about our day's hunt. Knots of muscle could be seen working over his cheek-bones. He stared at me in such a way that I'd rather he had hit me. A deathly silence fell on the tent. One of the dogs could be heard outside, gnawing at a bone.

'See how much they've caught for you,' Ignat said slowly to Kalugin, who was taking off his boots.

How could I answer his reproach? I was clearly in the wrong. To break the awkward silence, Kalugin tried changing the subject, and talked about the wild pig which he had injured earlier in the day. But Ignat remained stubbornly silent, and stared at the glowing stove.

About half an hour passed. Petrushenko pulled his watch from his pocket and exclaimed, 'What are we upset about? We'll see how things stand in the morning. Now we must see in the New Year.' Hastily he began unfastening his bag. Kalugin followed suit, and also pulled out roasted meat, suet[4] and honey. At exactly midnight we raised our glasses.

'Tiger on a Poplar'

January 1st turned out to be clear and calm. The light reflected from the snow was blinding, and the mountains in the distance were violet. We rose late. As soon as the tent was taken down we packed the horse's load. Between 11 and 12 o'clock, without stopping for tea, we left the sledge on the ice, carefully turned upside down by Ignat, and set off. It was decided to brew tea later, somewhere along the rapids.

Following the winding river we stamped out a path in the snow. The horse came along behind with Zabaluyev. Mist was forming over the rapids, and we had to detour around them along the bank. In the frosty air lingered the heavy scent of conifer needles. We were passing a burnt hillside. Zabaluyev had fallen behind with the horse and was out of sight. The way ahead was blocked by a fallen tree and Ignat decided to go around it, but just then came a shout from Zabaluyev. I threw off my pack, gave Petrushenko Dik's lead, and ran along the river.

I had not gone two hundred metres when I was stopped short by hearing a resounding shot behind me. Then a second, and a third.... Shots rang out one after the other. There was a tremendous racket going on. The choking bark of a dog reached my ears.

I rushed back and saw Ignat running along the river. Behind him was Petrushenko. The others were holding *rogulina*s[5] - poles with two branches forming 'horns' at the end. Only now did I recognise yesterday's bend in the river!

Kalugin came up and thrust a *rogulina* into my hands. I ran towards the barking dogs. In a thicket, about ten metres from the bank, I saw Petrushenko calling me desperately. Ignat was bending down, trying to see something. A branch cracked and Kalugin ran up behind. Ignat straightened up and turned towards us with tense features.

'The thicket!'

He pulled his collar up with his free hand, and crawled into the snow-covered branches of a fallen tree. We formed a semicircle around him, with *rogulina*s at the ready. It seemed an age before I spotted the tiger. It was standing on a fallen poplar trunk, its rust-red side brilliant in the sun.

'Now where will it be off to?' I wondered, as I watched Ignat crawling towards the animal. My legs instinctively carried me forward behind him. I was not watching either Kalugin or Petrushenko, but could hear their unsteady breathing.

The tiger slowly turned its head, and with a roar lept onto Ignat. Just two paces from us, it forced him down into the snowdrift with its paws. From the snow momentarily appeared Ignat's bleeding hand, its fingers spread wide.

'Fire!' I immediately thought.

But to fire would have been dangerous: Ignat could have easily been hit. Dik seized the tiger by the neck, while Sharik and Volchok attacked from behind. Kalugin and Petrushenko, their faces contorted, had applied their *rogulinas* to the tiger and, with all their might, were straining to force it off Ignat. Then I swung the barrel of my carbine and struck the tiger's snarling muzzle. It tried to get away and dragged the dogs along the snow. I threw aside the broken rifle and rushed over to help Kalugin and Petrushenko. Out of the corner of my eye I saw Ignat get to his feet.

The tiger roared, twisted and squirmed, and seized the *rogulinas* in its teeth, trying to extricate itself from them. With frozen hands we withstood its frenzied efforts. In ten minutes - which seemed like an eternity - the tiger's strength ebbed away. It fell limp and, ceasing its chilling roars, calmly rolled its eyes, which protruded from their sockets.

'Pull off the dogs', croaked Ignat.

In another ten minutes we had bound the animal's paws. It was a tigress weighing a hundred kilogrammes.

Exhausted, I leaned against a tree. My whole body was wracked with a dull ache, as if it had been beaten with sticks. I wasn't even up to talking. Kalugin and Petrushenko were sitting nearby. The dogs were sleeping, their muzzles buried in their paws. Ignat, still in a bad mood, was standing bareheaded and bandaging the wound on his hand using straps for tying up tigers.

After a rest we dragged our catch to the river. Here it could be seen that, about two hundred metres downstream, haze was rising. The horse was standing near the bank.

Ignat hurried away. He wanted to check both banks of the river to discover which way the other tigers were going.

We decided to brew tea immediately and recover our strength. Taking advantage of the opportunity I wanted to treat Ignat's wound with penicillin, but it was now discovered that Zabaluyev had lost my rucksack. It contained the cooking utensils, camera, first-aid kit and my notebook. Ignat lost his temper. All his pent up anger and frustration of the last few days were vented on Zabaluyev.

It was an unlikely scene: the tiger was roaring and writhing on the ground, the tethered dogs were barking, Petrushenko was holding the snorting horse, and Ignat was shouting, beside himself with rage.

'While we're catching tigers you can't even look after a saucepan,' he growled at Zabaluyev.

Kalugin and I tried to calm him, but he wouldn't listen.

'You call yourselves hunters! And you let our dogs get killed!' - he turned on me, referring to the events of the day before. Zabaluyev then started trying to justify himself, instead of keeping quiet. Ignat now really let us have it, and finished by ordering us to go out and find the lost bag.

Zabaluyev said that he had already searched as far as the River Kalinich, and had not found it. Ignat, seizing his rifle, went off himself. The weather had been getting worse. The sky was overcast, and the taiga was gloomy and no longer silent. I couldn't remain in the tent twiddling my thumbs and - asking to borrow Petrushenko's rifle, since mine was now in two pieces - I caught up with Ignat above the fallen tree. He still had not calmed down.

'Ask him to do the simplest thing, and he goes and loses the bags....,' muttered Ignat in frustration at Zabaluyev, but suddenly stopped short: here was fresh tiger spoor crossing our path!

'The tigress and just one cub, apparently,' said Ignat, loading his rifle.

The tigress, revealing itself by the missing toe, had been coming along behind, stepping in the cub's spoor.

'You look for the bag, and I'll go on ahead,' said Ignat and, stooping low, he crept into the riverside undergrowth. I continued along the path following the horse's hoof marks and, arriving at the sledge, could not find the rucksack anywhere.

'The devil take him!', I railed inwardly at Zabaluyev. The possibility that the bag might not be found within three kilometres hadn't entered my head. I glanced at the sky, which was ready at any moment to cover everything in fresh snow, and I became overwhelmed with frustration. To think, just losing the saucepans would mean abandoning the hunt.

After searching for a long time I was groping around in the snow under a steep clay bank and at last found the bag. It had evidently fallen from the horse when Zabaluyev was working his way around the fallen tree blocking the stream.

Meanwhile Kalugin and Zabaluyev had put up the tent. Not far off could be heard the sound of sawing. Zabaluyev was sitting by the scarcely flickering fire, looking aggrieved.

'We need to cut them off quickly' said Ignat, coming up with a log on his shoulder. His hunch was that the other cub had not managed to get far away yet. It had to be caught the same day. So we hastily prepared straps for binding the tiger, and set off along the river with the dogs on leads.

Along the bank in open places we came across about a dozen tiger lairs. At several it could be seen that the animals had played games after feeding. Walking in a large semicircle we found a single tiger track. It climbed up out of the narrow valley and then returned to the stream. The spoor was especially clear where the snow cover was thin. Ignat measured the heel: it was two centimetres larger than that of the captured immature tigress. We stopped and thought about it. The dogs sniffed at the spoor, yelped, and pulled us towards the burnt slope. The tiger was nearby. Taking his rifle from his shoulder, Ignat wanted to release Sharik.

'Tomorrow, perhaps? It would be better with four,' suggested Kalugin.

'But the tiger will be off during the night!', objected Ignat, still surly. He was interrupted by a shot from the direction of the tent. Two more quickly followed. Faint shouts could be heard. Imagining something dreadful happening my heart was racing. We ran back. The shouts became heart-rending. While running we let the dogs off the leash. Ignat raced ahead.

Running into camp I was met with a chaotic scene. Ignat was sitting on the tiger's back while Petrushenko and Zabaluyev were pinning it down with some stakes. On all sides the dogs were fastened on the tiger. Its front legs were free, and it was no easy matter tying them together again.

Until late at night we were sawing and bringing in logs for a cage. It was essential to untie the tiger as soon as possible and get it into the cage. Ignat feared that its bound paws would catch frostbite. We at last released the tiger into the cage between 1 and 2 o'clock in the morning. In the frosty air its roar resounded along the snow-covered valley. Exhausted and still wearing our outdoor clothes and frozen elkhide boots we dropped onto the conifer branches lining the floor of the tent.

In front of me my diary contains the following short entry: '2nd January: upper reaches of Khanikheza; burnt areas, tiger tracked, heel 9 cm; 2 lairs; it went in a large semicircle.'

These lines, written in the flickering light of a 'Siberian stove',[6] break off abruptly.

My memory of it is vivid. On that day we covered about forty kilometres, scrambling up ridges and dropping down into deep valleys. The tiger made its way through the taiga by routes known only to itself, paying no attention at all to other animals' tracks. It left the burnt area, crossed the small streams forming the upper reaches of the Khanikheza, and dived into dense conifer-mixed forest.

Towards evening the dogs began straining at their leashes. We followed the track up a steep ridge and came across a tiger's lair. The animal evidently heard us coming and took cover in the undergrowth. On account of the silence that had fallen on the taiga Ignat decided not to release the dogs.

During the night I was woken by the penetrating cold. I had cramp in my legs, and my teeth were chattering. There was a dull wintery sound coming from the taiga. The 'Siberian stove', on the point of going out, was producing a pungent smoke.

'Oo-hoo-oo-oo....Oo-hoo-oo-oo...', called a hungry owl incessantly through the snow-bound forest.

Between 5 and 6 o'clock Ignat woke up. He looked at me, at the starry sky, grinned for some reason and began to fill the saucepans with snow. Kalugin also got up. He shivered, lit a cigarette and, to no-one in particular, said, 'I had a dream. Ignat was running away from me, with a grin on his face. What does that mean?'

'What do you think? You sleep and you dream. Today we must catch the tiger,' he replied calmly, wrapping his feet in *portyankas*.[7] The dogs were standing around the hot logs, their tails lowered. They growled at one another while fixing their hungry eyes on the saucepan, in which their broth was boiling.

'Shush, you devils!' shouted Ignat. Suddenly Volchok jumped up and seized Dik by the throat. Jaws snapped and fur flew.

'Grab their tails! Their tails!' shouted Ignat, and we followed his example to separate the crazed animals.

'Not just the tigers would hear that, but a bear in its den ten versts away!' cursed Kalugin. Dik yelped and licked his bitten paw.

A frosty dawn was breaking when we took the tent down. Ignat went ahead. He walked stooping, knocking snow off branches and skirting around fallen trees. Conversation dried up. The taiga is not the best place for it.

The tiger track was a deep furrow, all day long winding this way and that over mountains and along river valleys, and was now climbing up a steep slope. It was past midday.

'Climbing a steeple would be easier,' remarked Ignat, taking off his pack and in the process knocking a mantle of snow off some bushes. Hunchbacked mountains bristling with trees rose into the wintery sky. The bright ribbon of ice on the Khanikheza shone in the distance, and the cliffs along its banks looked tiny from up here. Somewhere far below, three days journey away behind the mountains, lay hidden the taiga village.

It was now 4 o'clock in the afternoon. In an hour the frost would begin to bite at our noses and freeze our cheeks. The fur on the flaps of our fur hats would collect frost, while the clothing on our chests would be covered in ice crystals from our breath.

Evening fell. We surmounted the 'steeple', but Ignat just kept going. Eventually he stopped and took off his backpack. Kalugin and Petrushenko started preparing our bivouac. Ignat went over to a dead pine nearby and in the frosty air the saw sang as its teeth ripped into the wood. The dried up giant swayed and groaned, raising a cloud of snow. We were sawing for about an hour and a half. It sapped our remaining energy.

While we were getting our breath back I leant against a tree, gazed at the twinkling stars, and wondered how many more nights like this stretched before us. At these moments it seemed to me that the whole universe consisted of nothing but the cold starry heaven and endless mountains. The stinging frost made me stir. We set to work again in silence. Under the sawn off trunk we inserted levers and rolled it to a cleared space. Kalugin chopped off the sapwood with an axe and drove in wedges.

A second log was rolled up, and our Siberian 'stove' was ready. The fire crackled, and with flames engulfing the logs the 'stove' sprang into life. What bliss to stretch out on a bed of prickly conifer branches after a day constantly on the go!

'Second Tiger'

'Something has been killed,' said Ignat, scattering the snow with his foot and pointing to a piece of skin.

'Wild pig', suggested Kalugin.

It was a steep ascent. Our hearts were pounding and our knees shaking. Ignat pointed to the other side of the valley.

'Above the Solontsovy.[8] It is sticking to the burnt area again.'

The tiger track, proceeding in a gradual curve, passed along cliffs and once more approached the Khanikheza. Somewhere down below was our camp. Between trees on the other side, across a narrow strip of bog, could be seen burnt hills.

Had Toeless come to the Khanikheza or not? We'd been puzzling over this ever since the previous evening. Now we were expecting to find her spoor any moment.

'She's leading us away from it,' said Petrushenko with some disappointment, throwing away a cigarette end. Kalugin agreed.

'Maybe,' said Ignat, glancing up into a pine where a squirrel, out of sight, was showering us with snow. 'We must catch her up.'

Once again Ignat pressed on ahead, ploughing through the snow. Behind him, from one thicket to the next, jumped Sharik. Winter days are short. It was already afternoon when we came to the Khanikheza, and the trees' shadows had already grown long. The tiger track crossed over and entered the burnt area. At the top of the hill, among spiny *Aralia* bushes and young fir trees, Dik suddenly dashed forward and began sniffing the snow. Sharik raised his muzzle and noisily sniffed the air.

'Tiger!' exclaimed Ignat, in a strained voice, as he feverishly made ready to unleash Sharik, who was straining forward.

The tiger, which had been in a thicket, went straight down the steep slope to the stream. Suddenly it stopped and snarled. The cliffs seemed to draw in closer. A chill ran through our bodies. The tiger jumped off its rock and disappeared into the undergrowth.

'Off you go!' shouted Ignat as he released Sharik.

Petrushenko unleashed Volchok, and I Dik. Ignat pointed his rifle up high and fired. Kalugin did the same. I pressed on down, not noticing that the shots were continuing, one after another. They roared and echoed, exciting the dogs flying towards the tiger, while the smell of powder filled the air.

Ignat raised his hand, and the fusilade immediately ceased. There was a tense wait - one minute..., two minutes.... Then the dogs' barking shattered the silence.

Setting off at a run Ignat suddenly pulled up short, pushing aside a branch. Its hard point had jabbed him in the face. Clutching his eye and uttering a short groan he began to sink to his knees. The burnt branch that had injured him was still waving back and forth. We crowded around. The left side of his face was covered in blood, and we could not see whether his eye was damaged. Kalugin and Petrushenko raised Ignat to his feet, and feverishly searched for a bandage in a rucksack. The barking of the dogs and the resulting echoes filled our ears with an incessant ringing sound.

'Over to the tiger! We'll lose the dogs!' shouted Ignat, as he gathered his wits.

I ran after Kalugin, leaving bits of my coat on *Aralia* spines. The fight with the tiger was taking place in a small patch of burnt spruce. We dashed around looking for branched poles to serve as *rogulina*s, but there wasn't an oak or birch in sight.

'What shall we do? What shall we do?' shrieked Petrushenko as he ran around with an axe. Then I caught sight of Ignat. Covering his eye with his hand, he was forcing his way through a burnt thicket, brandishing his rifle. He shouted something.

'No *rogulina*s!' replied Kalugina.

'Cut spruce!' screamed Ignat. Running up, he snatched Petrushenko's axe and began cutting.

With branched stakes clutched in our hands we took up positions near the dogs. They were running at and jumping back from a large log with snow drifted up against it. Behind it could be seen the tiger's head, with its fangs bared. It made a mighty leap, snapping off a small dead tree in the process.

What happened was obscured in a cloud of snow. Sharik was flung aside like a reddish tangled ball. The tiger snarled. When a gust of wind brought it our scent, it jerked its head around and, seeing us, opened its mouth even wider.

Ignat shouted a warning, but the tiger had already launched itself towards us. We caught it on our stakes, as if they were bayonets. Then came a blow from a paw! A pain shot through my arms and our stakes flew through the air. The animal's stinking breath overwhelmed me, and a paw caught my coat. I quickly snatched up my broken stake.

The tiger crunched up the ends of the stakes, and reared up on its hind legs. Turning aside, it dived into the snow with a roar. Ignat and Kalugin, grabbing a pole, one at each end, pressed it down onto the tiger's chest. There was roaring, shouting, barking, yelping, snarling and growling! Ignat was holding his end of the stake down with his knees, and with his hand, from which dangled the binding straps, touched my shoulder and shouted something. At last I realised what he meant, and took over his end of the stake.

Ignat jumped up and, using all his strength, managed to tie the tiger's hind paws. It lept into life with renewed energy, roared even more loudly, and began to force up the stake holding it down. Ignat pulled out of his trouser pocket a twisted length of string and hastily made a loop. At the second attempt he managed to throw it over one of the tiger's front paws, and pulled tight. It was another five minutes before the paws were secured.

We sat in the snow and used our hats to wipe away the sweat pouring down our faces. I examined Ignat's wound, breathing a sigh of relief: the eye was still in one piece. The sharp tip of the burnt stick had sliced the eyelid and pierced the eyebrow.

Hardly able to stagger along, Kalugin and Petrushenko went off to fetch the rucksacks. The dogs licked their wounds. It was getting dark. The sky was covered in leaden cloud. Ignat, slouched and with his face in his hands, was sitting on some branches. We were shattered. With vacant expressions we gazed at the roaring tiger. We decided to get it back to camp.

'Even if it takes all night. And then we must make a cage.' Wincing with pain Ignat began hoisting his rucksack onto his back.

Once the burnt area was crossed, we dragged the tiger onto the stream ice. It was snowing heavily. Kalugin went ahead. Ignat took up the rear, grumbling when we were so tired that we had to stop for a rest.

That night long remained in our memories. We had to force our way across snowdrifts on the river, each time sliding down the far side with the tiger in tow. If the river was blocked with fallen trees we had to force a way through thickets on the bank, where the snow was waist-high.

Suddenly I noticed that Dik was not with us. We stopped and began calling, but the shouts were lost in the sound of the wind. I went back along our track. Shattering the surrounding darkness with rifle shots I slowly made my way, calling the dog. Unexpectedly I felt something soft under my feet. It was Dik. He lay in the middle of the path. I bent down and he, yelping excitedly, licked my face with his rough warm tongue. The dog was worn out and could not carry on. I had to drag him along on a leash.

Joining the others we staggered on for another three hours. No-one had any idea how far it was to camp.

'Ignat, maybe we should find a place to bivouac?' suggested Kalugin.

'How wonderful, to be in our tent right now', I thought. The tiger began to struggle and growl deeply. Then we heard answering shouts, and shots came from close by.

'Camp!' exclaimed Petrushenko.

Gathering our remaining strength, we dragged the tiger to the tent. The frightened Zabaluyev told us how the whole day he had been pestered by a tigress roaring and circling the camp through the surrounding thickets. To scare it off he had used up all the ammunition.

By the light of tar flares we moved apart the planks from above and got the tiger into the cage. Through gaps in the bars we untied the straps. The tigers immediately stopped roaring and began licking each other's face.

Zabaluyev stayed up to guard the camp, while we lay down to sleep. But sleep for me was impossible: my body ached, my head was buzzing, and my mind was filled with tiger tracks. They flickered, disappeared, reappeared, and again flickered before my eyes. With a superhuman effort I banished this vision, but then a tiger baring its fangs took its place: it came nearer and nearer, growing to a colossal size. I started up, came to my senses, and once more threw my coat over my head.

I woke up exhausted. It was already light in the tent. Beside me I thought I recognised Kalugin, but the face was different, with prominent cheek-bones. On the other side of the tent was another newcomer, in a short quilted jacket. I went outside, and the first thing I saw was the cage with the two tigers. Zabaluyev was preparing food at the fire. I asked him about the strangers.

'Hunters. Got lost during the night. Because of them, after midnight we had to sleep sitting up.'

About two hours later we said good-bye to the hunters. They were Ivashchenko and Muralevich from Ariadnoye village. They reached the Khanikheza by tracking a bear they had wounded fifty kilometres away.

During the day we constructed two long deep cages. By evening Kalugin had returned from hunting. Ignat began feeding the tigers. He pushed meat on a forked stick through the bars. At first they snarled and pawed the meat, but soon fell to and ate greedily.

Next morning, at dawn, we transferred the tigers into the new cages and loaded them onto sledges. Water had collected on the river ice, and mist was rising from it. It meant that we could not reach Malinovo before nightfall. We pitched the tent and spent the night on the way.

The following day was sunny. Trampling a path through the snow for the horse, we saw in the distance a man coming our way along the river.

Two hundred paces..., one hundred..., and we ran towards each other. Before us stood the smiling Makar. He was wearing a new coat, new cloth trousers, and new boots.

He took off his pack and straightened up his shoulders.

'So I missed the fun, eh?' said Makar, slapping Kalugin on the shoulder. 'Well, I'm in no hurry to go anywhere. In November we brought in the soya, then came New Year.' And Makar, staring at the bandaged Ignat, cheerfully inquired, 'Well, brother, what happened to you?'

We unharnessed the horse. The fire was soon burning brightly, and saucepans appeared over it. Makar thrust his hand into his pocket for tobacco. As well as his baccy pouch he brought out a piece of paper.

'I'm forgetting the most important thing. I'm in a muddle,' and he handed over a telegram. We were instructed to return immediately. Plans had changed. Approaching Malinovo, the team was to go after a third tiger.

Ignat remained silent. At last he said, 'There is nothing to do at Malinovo. You can take the tigers yourselves.' He pointed to the blue hills in the distance. 'We'll spend the night at Kisly Stream, by evening we'll cross the Kalinich. Another night under the pine tree, then the burnt stretch. It is hiding out there.'

Although Petrushenko was hoping to spend a couple of days at home, Kalugin and Makar dissuaded him. We drank a mug of tea and began to get ready. Makar tied Dik to his belt.

Ignat set off across the river towards the steep bank opposite, and in about three minutes they had all climbed up it. Against a background of snow-mantled bushes the figure of Ignat showed clearly, with the long rifle over his shoulder and Sharik on a leash. Bent under the weight of his heavy backpack followed Makar, and the tall lean Kalugin took up the rear. Petrushenko stayed behind a while, before catching them up.

At the next bend in the river Ignat stopped, turned and waved, before disappearing into the forest. One by one they were lost to view. It was a sad moment. In a flash I recalled all the past days of joy, disappointment and privation. I still saw before me Ignat's gaunt face with its determined mouth.

'Best of luck, my friends! Return soon!' I said to myself, bidding them farewell until our next expedition.

Cover of Lev Kaplanov's book Tigr izyubr los' (Tiger, Izyubr, Elk) *published in 1948 by the Moscow Society of Naturalists*

Chapter 13

PEOPLE, POPULATION, POACHING

The first census of Amur tigers was undertaken by Lev Kaplanov in the winter seasons of 1939-1940 and 1940-1941. He estimated that only 25-30 survived in the whole Russian Far East.[1]

The Amur tiger was subsequently rescued from the brink of extinction largely through the efforts of the Soviet scientific community, especially the Sikhote-Alin Zapovednik scientists K.G. Abramov, Yu. Salmin and Kaplanov himself. Following Kaplanov's tragic death other tiger experts, particularly Abramov, urged the authorities to save the tiger. In 1947 it was granted legal protection in the Soviet Union, and hunting by shooting, poisoning or trapping was banned. This was the first time that a tiger population anywhere in the world had been protected in this way. Legislation for tigers in other countries was not brought in until the 1970s.

The 1947 ban was too late to save the other tiger found within the borders of the Soviet Union - the Caspian tiger. Its main population was destroyed in the 1930s as reed beds along the great rivers of Central Asia were cleared for cotton cultivation. Although it had disappeared from the Soviet Union by the 1950s a few survived in northern Iran into the 1970s.[2]

In the Russian Far East exceptions were made for tigers attacking domestic animals or people, though attacks on people were practically unknown at this time. After 1947 some 2-3 tigers were legally shot each year as problem animals, or in a few cases they were moved to other more remote areas.

Poaching of tigers did take place occasionally; the victims were usually young, inexperienced and less secretive animals. There was little motivation to hunt tigers as no real market for tiger products existed in Soviet times. However, hunters sometimes shot tigers because they saw them as competitors for deer and wild pigs, and others were killed by villagers who felt that they or their domestic stock were threatened. The carcasses of illegally killed tigers were often burnt or buried as there was no way of selling their skins or body parts, and punishments could be severe.

Tiger numbers increased in the Russian Far East from the 1950s.[3] The recovery was at first quite slow as reduction in the number of hoofed mammals by human hunters led to shortage of prey for tigers in some areas. In 1956 the live capture of cubs was banned for five years, and licensed afterwards, though these measures had in practice little effect. Many cubs were still being captured each year for zoos and circuses, this being more profitable than shooting tigers for their skins.[4] Four capture sites were used - two in Khabarovsk Region, and two in Primore - and several tiger-catching families operated from these centres.[5] Some 32 cubs were captured between 1947 and 1956, and about 53 from 1967 to 1986. Between 1920 and 1960 as many tigers were captured as were killed by hunters.[6] As a result, in 1964 the tracks of only two wandering tigers were recorded in Sikhote-Alin Zapovednik, with no permanent residents.

Did live capture of Amur tigers cause long-term damage? The question is not easy to answer, but even if it did, the removal of so many young tigers from the wild ultimately resulted in a large and well managed captive population - a necessary insurance policy for the great striped cat's survival.

When in the 1970s protection from hunting was reasonably well enforced, the Amur tiger population grew significantly. In Khabarovsk Region, as hunting by humans for meat and fur increased and the road network for transporting timber was extended, the behaviour of the tigers changed. Whereas previously tigers avoided people, increasingly in the 1970s and '80s they simply would not give way to people or to vehicles. They were recorded wandering right into villages or bee-keeping settlements to capture dogs or terrorise villagers themselves. Attacks on farm animals increased as the tigers sought substitutes for their diminishing wild prey.[7] Dozens of hunting dogs were devoured, traps for fur-bearing animals were ruined, the meat of shot game was carried off and hunters' huts were boldly approached. As a result local people felt more hostile towards the tigers.

Nevertheless the numbers of tigers in the Russian Far East continued to rise and by 1989 it was estimated that there were some 430 - more than at any time since the First World War. This represents an approximate halving of Russia's Amur tiger population compared with the mid 19th century. In India, by contrast, where the largest tiger population survives, the 3000 or so living in the wild today are only a tenth of the 19th century population of 30,000.

Other authorities, however, put the number of Amur tigers in the 1980s at nearer 250, suggesting that higher estimates were the result of the nomadic propensities of the great cat.[8] Interestingly a similar range of estimates was produced in the 1990s: according to WWF, based on winter tracking, there were 415-475 tigers in the Russian Far East in the winter of 1995-96, whereas a study in Lazovski Zapovednik at about the same time, using sniffer dogs, suggested that the Russian Far East contained only about 250.[9] Evidently counting tigers is not yet an exact science, and it is difficult to reconcile the apparently large tiger population with the supposedly deteriorating ecological conditions for the tigers.

In northeast China it was estimated that there were about 200 tigers in the early 1950s.[10] While in the Russian Far East protection of tigers has had some success and a great deal of field research has been carried out, in China little was achieved. Tigers were reduced in number to the brink of extinction by the 1980s and '90s, as a result of hunting and deforestation leading to loss of prey and habitat. The top-grade protection announced in 1982 has not been enforced. In Korea tigers were abundant throughout the peninsula in the 18th and early 19th centuries, but by the early 20th century they were common only in the north; in the 1990s few if any remained, and those only in forested mountains near the Chinese border.[11]

Felling of forest

Outside zapovedniks and other protected territories most forest areas within the range of the Amur tiger have been cut over as many as three times since the early 20th century. In local communities, logging provides essential firewood for heating and cooking, and timber is needed for construction. Timber extraction has traditionally been an important part of the regional economy and a large employer of labour.

From the end of the Second World War until the 1980s the local timber industry was based on Korean pine. It was the most valuable commercial tree in the region, and its kernels provided valuable food for wildlife. By the mid 1980s the species was clearly being over-harvested, and in 1990 the Primore administration outlawed felling of the pine except under special circumstances. In the chaotic state of Russia in the early 1990s, however, the export of Korean pine continued largely uncontrolled. By the mid 1990s most of the saleable pine had already been removed and logging was mainly of other species.[12]

The desperate need for foreign exchange has understandably compelled the Russian authorities to seek co-operative ventures with international corporations that can invest capital for large-scale projects. Both legal and illegal timber operations have been attracted to the former Soviet Union, and modern technology is more damaging to the environment than previous Soviet neglect. Japan, South Korea and China are major markets. Felling capacity has already reached unsustainable levels in Primore, where most of the tigers are found. While clear felling is practised in Khabarovsk Region along the northern border of the tigers' range, in most of its range selective cuts are made to extract just the highest quality timber from the mixed forests.

The effect of the logging industry on the tigers' habitat is complex and not always entirely negative. After areas have been cut over, new growth of herbage and other plants attract deer, and then tigers come to prey on them. Wild pigs, on the other hand, prefer mature Korean-pine or Mongolian-oak forests, where they feed on kernels and acorns, and the felling of these forests may be a major reason for the pigs' steep decline.

More insidious effects may ultimately have a greater impact on tigers and their prey than logging itself. In contrast to clear felling, selective logging requires work to be conducted over large areas. So an extensive road network is developed, and this provides access for hunters. As a result more animals of various species are killed, both legally and illegally, and this brings a greater chance of tigers being poached.

In the eastern part of Sikhote-Alin Zapovednik, which is cut by a permanent road, only 2 out of 14 tiger cubs born from 5 litters between 1991 and 2000 survived to disperse. Increasing numbers of tigers have been killed by timber lorries or cubs have died after their mothers have been poached by hunters using the roads for access. Traffic disturbance causes tigers to abandon their kills near main roads, so increasing the time needed for hunting, reducing their periods of rest, and leaving them exhausted.[13] Away from the main roads however tigers are surviving well. It has been suggested that timber roads should be closed off once logging operations are completed, and that other roads giving hunters access to sensitive wildlife habitats should be closed or have guard-posts or barriers erected.[12] Where this is being carried out the hoofed animals crucial to the tigers' survival have an opportunity to recover.

Local people desperately need employment in forestry, and timber extraction is bound to continue; but if this can be managed in such a way as not to destroy the local wildlife, some measure of co-existence between tigers, local people and economic development may be possible. To reduce the incentive for local people to hunt wildlife other economic activities must be available. The encouragement of woodworking enterprises - such as making furniture, doors and other products - offers employment, utilises a local sustainable resource, and keeps value-added profits within the region. With Russia's economy and its wood-products industry on the rebound after the 1990s - when timber harvesting fell by 60% - regional Russian enterprises have begun processing timber into chipboard for export.

Forest fires

Forest fires occur frequently in the Russian Far East and many adversely affect the Amur tigers' habitat. In the extreme southwest of Primore, as much as 40% of Khasanski District is burnt over each year to rejuvenate pasture. These fires often merge into a vast raging inferno devastating wildlife habitats.

Early summer 1998 was very dry in the Russian Far East, bringing a high risk of forest fires. Fires did break out in Sikhote-Alin and Lazovski zapovedniks in May: 3000-4000 hectares were destroyed in each reserve, including valuable Korean pine forests. One adverse result of perestroika in the late '80s had been the disbanding of the Russian federal forest and environment service, and lack of fire-fighting equipment and adequate fire-breaks presented serious problems during this crisis. Foreign organisations immediately sent funds, and by mid-June, with welcome rain falling, the fires had been put out by the zapovednik workers.[14] There were more forest fires in 1999, 2001 and 2002.

These terrible fires are caused by lightning, by careless picnickers - many forests being accessible by road from urban areas such as Vladivostok and Ussuriysk - and especially by wasteful logging practices that involve leaving behind vast quantities of dry wood, sometimes 40% of the timber cut.

The fires have been very serious indeed for the timber industry, wildlife and the region as a whole. Apart from destruction of forests and the deaths of many animals, others have been driven into settled areas, where they can be more easily shot. Although tigers can outrun such fires, they may find themselves in other tigers' territories - with consequent conflict between the animals.

In the long run however these fires often encourage new undergrowth, which provides fresh fodder for hoofed animals. By increasing the number of prey, this can be to the advantage of the tiger - as happens after regular burning in the national parks of Nepal and India.

Tigers and people

During the 1980s the Amur tigers' fortunes were mixed. In some respects things were going well: tigers had increased in number at a greater rate than at any time probably since the early 1900s. On the other hand the abundance of prey had been greatly reduced by excessive human hunting and adverse natural conditions. In the winters of 1982 to 1984, and again in 1988, heavy snowfall and disease reduced the number of hoofed animals and there were failures of acorn crops because of disease in the oak trees. After high mortality of wild pigs in the severe winter of 1983-84, when many were killed by tigers, there was more competition between tigers and human hunters in the following winter, as the pigs had produced few young in the intervening season. 'Tiger herds' - groups of up to seven, usually sub-adults travelling together - were observed in Lazovski Zapovednik by Vladimir Medvedev. This seems remarkable for an animal usually described as solitary, but cubs do stay together for a long time, and Medvedev's observation may reflect both an increase in tiger numbers and the scarcity of prey. Despite some poaching of tigers, years of protection may have caused them to relax their caution towards people. This was probably true particularly of younger and less experienced animals soon after becoming independent. Some may have been seeking fresh territories or, at a time of growing tiger numbers and diminishing prey, they may simply have been looking for food.

In the late 1980s the appearance of tigers in populated areas was enough to cause something of a panic. Between 1984 and '86 two healthy adult tigers were noted around Vladivostok, occasionally even coming into the town. They lived almost exclusively on feral dogs and were rarely seen by people. However, in 1986 one of these tigers - a young male that had been preying on dogs - was located sitting near a tram stop. The two tigers were later shot.[15] In the next couple of years people were sometimes delayed in their journey to work by tigers sprawled across the railway line, and they even appeared at the windows of ground-floor apartments in 1987.[16] In another case one was found asleep in the hallway of a Khabarovsk flat; it was captured and returned to the wild.[17]

There continued to be many tiger attacks on domestic animals, including cattle that often graze freely in large herds around villages. More serious for the tigers' image have been human fatalities. Whereas there had been no cases of man-eating tigers in the Russian Far East since the First World War, in the 1970s there were renewed instances of Amur tigers attacking people. In 1976 a tractor driver near Lazo was killed and eaten by a tiger. Later there were several other similar cases.[18] Young tigers were probably becoming bolder, though in the majority of instances the human victims were pursuing the tiger or its prey. Cornered or wounded tigers are dangerous.

At a lodge of a hunting inspector, whom I visited in August 1992 on the Keya River, a tributary of the Khor, barriers had to be erected around the kennels of hunting dogs. A tiger had killed no less than eight in quick succession, only one of which was eaten. The inspector fired warning shots to drive the tiger away, but it kept returning. Even so he seemed rather well disposed towards the tigers and the beautiful forests that surrounded his home.

Others, however, are not so sympathetic to the great striped cats. Hunters in the Bikin Valley alleged that tigresses teaching their cubs to hunt sometimes kill whole families of wild pigs, but may eat only one animal. When there is a heavy snowfall the wild pigs' trails become long trenches, almost like tunnels, and if they are trapped in these by tigers a great slaughter may result, leaving fewer for human hunters. One villager said that he had several tigers in his hunting concession, and claimed that they take a heavy toll of wild pigs.

Russian tiger experts were faced with the growing conflict between people and tigers in the mid 1980s, for although the overall human population was not growing significantly the use of natural resources was intensifying. One proposal was that the number of tigers in southern Khabarovsk Region should be reduced from some 50 animals to about 25. This could be done by trapping adult tigers in areas where they were numerous. Then two or three populations of Amur tigers could be established in other suitable parts of the Russian Far East using the captured tigers. It was argued that the tigers would survive in these places, as they were tough enough to live in any conditions provided there was sufficient prey. For instance a tiger reserve could be established on the left bank of the Amur, where they are no longer present, once it had been stocked with wild pig, which would be provided with soya-bean fields and other fodder. It was also suggested that the traditional methods of catching tiger cubs in the Russian Far East should be abandoned - methods that sometimes caused shock, injury and even death to the animals; immobilisation using dart guns would be safer for the animals and more effective.[19] Although these interesting and constructive proposals have been considered, little action seems to have been taken.

Other tiger experts began to question the whole strategy of tiger conservation in the Russian Far East, even suggesting that trophy hunting might be reintroduced.[20] There were proposals to reduce the total tiger population to a more manageable number of 200 or so within areas that could be more easily protected. Sport hunting of tigers would, it was argued, bring in valuable income for the leaseholders of hunting lands. However, these suggestions have not been implemented because of the alarming consequences of the Soviet Union's collapse in 1991.

Trade in tiger parts

Prior to the dissolution of the Soviet Union there was relatively little incentive to kill tigers for profit, though some were killed for other reasons. Under the Soviet system tigers and other rare wildlife were supposed to be protected, both in law and in practice, with strict controls on firearms and hunting activities. Contacts between Soviet citizens and foreign businessmen were limited as borders were closed. The only market for skins and other animal products was through a state organisation, and since tiger hunting was outlawed there was little opportunity to sell tiger parts.

However, with the collapse of Soviet power numerous state enterprises were disbanded. This left many people in the Russian Far East with no income. So, in the ferment of privatisation and commercialisation, wildlife became part of the market economy and tiger parts were substituted for lost jobs.

Loosening of political controls and the opening of borders made access to the market for traditional Chinese medicine much easier. The illegal market for tiger products became extensive, threatening the tiger as a species worldwide. The demand for tiger derivatives from Russia, as well as from India, rose sharply at a time when stocks of tiger bones in China were becoming exhausted. One kilogram of tiger bone, ground down as a medicine, can fetch $2000 in the Asian medical market.

The blood, meat, penis and whiskers of the tiger, as well as other parts, also command very high prices when sold for traditional Chinese medicine. At the open market in Khabarovsk a first-level middleman offered investigators a frozen tiger carcase for $5000. Being among the largest of the tigers, the street value of an Amur tiger's body can reach

$60,000, so that in the early 1990s the tiger became a major target for poachers, the Russian 'mafia', and international illegal traders.

Prices of skins depend on the size of the animal, the hunter's knowledge and access to the market, his readiness to store them until prices rise, and the season when it is killed - since the fur is in prime condition in winter. According to customs officers at Vladivostok port in 1994, a sailor was arrested in possession of a tiger skin he had purchased in Russia for $1000 but hoped to sell in South Korea for $3000. In some countries a tiger skin can sell - as a decoration - for $15,000.[21]

Sea-going vessels are the most popular means for exporting wildlife from Primore as cargo ships leave from several large and many small ports along its extensive coastline. Ships carrying timber are used to smuggle tiger parts, concealed between the logs. Wildlife products are also carried across the border into China by train or motor vehicle.

Most tiger bone from the Russian Far East goes to the Chinese and South Koreans. China's main interest seems to have been in processing and re-exporting tiger bones in the form of commercial medicines and treatments containing tiger derivatives - such as tiger-bone wine as a tonic, and plasters impregnated with tiger bone for treating rheumatism. On the other hand the majority of tiger pelts seized by Russian Customs officers have been bound for Japan.

Poaching

The dramatic increase in tiger poaching has been prompted by the new accessibility to major markets in China, Korea and Japan, the ease of acquiring firearms, and the increased numbers of people illegally hunting for game meat, which in turn increases the chances of tigers being found and shot. Since the early '90s therefore, when the borders opened to foreigners and international commerce, traders from China and South Korea have flooded into the Russian Far East in search of business opportunities.

Russian poachers responded quickly, using snowmobiles, motor vehicles, helicopters, horses and dogs to hunt down tigers, as well as bears and other wildlife. In some cases the tigers are caught with snares and traps, but mostly they are shot on chance encounter or after being tracked in the snow. Some are caught in traps set near the remains of prey or occasionally along tiger paths or by marked trees. Poachers often catch mothers with cubs or sub-adult tigers; the more experienced and cautious adult males are taken more rarely.

Despite the prohibition of tiger hunting, enforcement was lax during the unsettled early '90s because of inadequacies in funding, equipment and trustworthy personnel. The law courts did not seem to treat poaching as a serious offence, and it was often difficult to draw the line between legitimate hunting and poaching. Penalties, at a time of runaway inflation, were derisory. As a result poaching became widespread.

In the illegal trade in tiger parts there are two tiers of middlemen. First-level middlemen live in the villages, small towns or district centres, and have close contact with the local people. From them they buy forest products, including wildlife products such as tiger pelts or bones, often in exchange for vodka, wine, cigarettes, sugar or tinned meat. Second-level middlemen are usually businessmen living in the cities, especially large ports and border towns. They purchase products from first-level middlemen, or sometimes directly from hunters, and may place orders for particular items and quantities. They then export this contraband themselves or re-sell to persons or organisations with access to good export channels. It is difficult to assess what proportion of poached tigers is exported. There were reports that in 1993 a shipment of 500 kg of tiger bone was seized in Harbin - representing 16-20 tigers, assuming that the weight of a large tiger skeleton is 25-30 kg.

According to Dmitri Pikunov there were in the early 1990s about 25 localities where tigers survived in Primore and Khabarovsk Region. In every tiger area there lived no more than 2-4 professional hunters ready to kill a tiger to order. Therefore, he estimated, there were altogether 50-60 active poachers who might kill a total of 25-35 tigers each year.[22] Approximately the same number of tigers were killed near deer farms. In addition, poaching from vehicles led to the death of 5-10 tigers each year.

Chapter 14
TIGER MONITORING[1]

Counting tigers is no easy matter. Organising a survey over the whole of Primorski Region, as well as the southern parts of Khabarovsk and Amur regions, is time-consuming and expensive. It is much easier to make a count on a fixed monitoring area, for example in a nature reserve.

It was more than 20 years ago that annual estimates of tigers were started in Lazovski and Sikhote-Alin zapovedniks. The results show that the tiger population in Sikhote-Alin Zapovednik has stabilised at 15-20 adults with 2-14 cubs,[2] and 8-18 tigers in Lazovski Zapovednik. Nearly all the tigers also regularly wander outside the reserve boundaries.

Annual surveys in the reserves give an idea of long- and short-term changes in population size, as well as the tigers' movements and behaviour. The results can also indicate possible undesirable or unexpected changes in the whole tiger population of the southern Russian Far East.

In our Sikhote-Alin Zapovednik, systematic and comprehensive surveys began in 1971. Fourteen routes totaling about 500 km were planned through the main catchment areas. Twenty-eight members of staff go out simultaneously into the snow-covered taiga in order to count the animals.[3] I'll describe a typical survey trip.

Long before the appointed time, everyone knows that there is going to be a winter survey of animals. We all make preparations, knowing that it will be challenging and physically demanding.

Strictly speaking the tigers are actually being monitored all year round, and not just along the winter routes. For us, the tigers are special. All staff and visitors at the zapovednik, regardless of age and seniority, are expected to look out for and record any spoor or other signs of the great cat.

It would be strange indeed if you failed to notice a footprint the size of a saucer. Having found one your field notebook should acquire an entry along these lines: 'tiger spoor 14.5 x 16 cm, heel 12 cm, stride 70 cm', followed by date and location. Although this is a chance observation and not on a predetermined route, even this is a form of monitoring,.

In summer the spoor on its own does not tell us much, apart from the fact of the animal's presence. A summer spoor is normally only 2-3 footprints, giving us an idea of the animal's size and direction of movement, and the date when it passed.

But where did the tiger come from? Where was it heading? What did it see? The 'book of the taiga' has been slammed shut, and only a page fallen from it gives us some idea of what has been written in the book by the tiger. If only we could see what was written on the other pages....

Long and serious preparations lead up to the winter count. The routes are decided, the monitoring parties are selected, and then come the final instructions.

The survey covers a large part of the reserve, and all the observers go out on the same day. Fastening on our skis in the grey dawn light we already know that, also standing on their skis, are all the people with whom we were discussing, swopping ideas, and receiving instructions at the zapovednik headquarters the day before. Now everything depends on us - and on our rucksacks, the route, our field notebooks, and 'our' tigers.

We are in no doubt that we'll find tiger spoor, and that the blank pages of our notebooks will be filled by the end of the route. The prospect makes me impatient, and prompts me to glance at my companion. In no hurry he is still adjusting his skis, and painstakingly rolling up an odd piece of string before stuffing it into his pocket: it'll come in handy. And I myself, mentally or with a glance, am checking the things I'm taking. We must not forget anything essential, or even just possibly useful. I involuntarily reach for a length of strap, which I cast aside yesterday as unnecessary: it might just be needed.

'Well, everything OK?'

'Let's go!'

The first few metres of untouched snow glide past under the skis. Our eyes are dazzled by the play of colours from sparkles reflected off the snow. The trees seem to be sprinkled with hoar frost, and bushes are decked out in fabulous beads. For the first few kilometres we are still fresh, but - at the expense of some energy - we can't help being distracted by details of the surrounding taiga. Not for the first time we are struck by its unreal fairy-tale beauty. After a while we suddenly notice that our collars and hats are coated - like the taiga - in a layer of hoar frost, glistening in the sunshine. At the same time we begin to feel the weight of our rucksacks. And just at that moment I hear my companion's triumphant shout, 'Over here!'

Even from a distance I can see a deep furrow, with loose edges of turned snow. Time for a rest. Rucksacks come off, notebooks come out. On page 1 is already the date, route, a note of the weather, and the spoor already found - wild pig, roe and *izyubr*, and several squirrels and Siberian weasel.[4] And here, at last, is the spoor we've been especially looking forward to, even if with a touch of fear.

First of all, we note the spoor measurements - length, width, heel size, and stride length. How fresh is the spoor? In fresh deep snow it is easy to tell the tiger's direction of movement. Now we have to try to discern, from the track, where it came from, whether it was making a journey or hunting, and if the latter what was it after? So we follow the track, which means going back the way we had come. We know of course that the tiger has keen senses and is very wary; even so, we nurture the hope, in some corner of our consciousness, that perhaps, in the next moment or two, ...

Although we have traveled this route many times and are familiar with it, we are still curious to know what is around every twist and turn in the path.

Today's stage is one of the shortest on the route, but for all that not at all easy. The first day's exertion is making itself felt. The overloaded rucksacks are still at their heaviest, so the first ten kilometres of the route feel like all the rest put together.

In the evening, after reaching a hut, we check our notes - the number of tracks of squirrel, weasel, mountain and Manchurian hares;[5] on the river ice those of mink[6] and otters;[7] and among the hoofed animals - pig, and *izyubr* and roe deer. Altogether there weren't many.

My colleague's disappointment soon evaporates however: 'Never mind. We'll have more luck later. We're in birch and thin larch forest. Wait till we get to the pines!'

I too know that we'll find more life in the pine forest. Even so, I'm left with the feeling that I've missed something when examining the spoor, a curious feeling of failure in not finding anything out of the ordinary. I note down all these musings and also the firm intention to compare our results with previous years'.

After breakfast the next morning we quickly get ready and are soon once more on our way. What has today in store for us? Well, it begins as soon as we leave the hut. Just around the first bend in the path we come across fresh tiger spoor. The animal climbed up from one of the river channels and came down the path, instead of along the river. Here the ice on the river is full of holes and melted patches, so the tiger preferred the firm path to the deceptive smoothness of the ice. Sensing the proximity of human habitation it stood for a long time at the bend in the path, even lying down in the snow; then it returned to the ice and skirted around the cabin. Crossing the path we came along yesterday, it headed for the hill on the right bank.

Imagine our surprise when, at the very foot of the slope, we found three other tiger tracks of various degrees of freshness. The sizes were similar - 16 x 13, 15 x 12, 14 x 12 cm. Most likely the differences were due to different ages of the spoor: the oldest would have melted more. It looked like a tiger path. In any case, it was a route used repeatedly, and the tiger would probably come this way again. We swopped ideas about how permanent this track was, and agreed that we'd get the answer back at zapovednik headquarters, when everyone assembled and someone else would have noted the movements of 'our' tiger. The aim of the survey is to get a general picture of the movements and resting places of every tiger in the zapovednik. We had just one fragment of the whole mosaic.

For the next few kilometres of the route nothing unusual turned up. Squirrel tracks were commonest, with *izyubr* coming a close second. Now the path crossed the next valley with a stream roaring along deep under the ice. The track of a large brown bear stretched straight along the ice over the stream, before joining the path.

'Good. Now we'll get along faster: it has beaten a path for us. And the spoor is fresh, not hardened. Let's hope it goes all the way to the hut.'

Indeed the going was now easier. We sped along, and as a result almost missed a tiger spoor - yet another one, and in this case along the path. It was following the fresh bear track! Finding the two together always promises to be interesting. Our spirits are lifted in anticipation, and there is a real prospect of seeing something exciting. This is soon confirmed when, around the next bend in the path, we pick up a large tuft of bear fur. A bit further - two more of the same. We keep our eyes skinned, almost bending down to sniff the snow, but find no traces of a fight.

The tracks - quite ordinary but of two massive predators - stretch uniformly through the pine forest, following the twists and turns of the path. I involuntarily find myself speeding up at the approach to each bend, but at the same time exercising due caution, all the time listening to the even sough of the wind in the tops of the pine trees. If the ear catches even a distant roar or growl, even though nothing is seen, we'll be satisfied: 'At last!' But the path carries on, turning this way and that, with the animal tracks likewise continuing ever on through the deep snow. Here there is a sharp turn in the path downhill, away from the contour we are traveling along and down onto the floodplain towards the hut, now not far off. We ought to turn and descend, along with the path, but the tracks of the two 'lords of the forest' carry on along the contour of the slope to the top of the next side valley. We stop for a brief smoke and an equally brief discussion: 'To the hut or after them?'

'After them, of course!'

Without our rucksacks and dispensing with conversation, we make rapid progress through the scrub. After a kilometre and a half there is nothing new, except for two more tufts of bear fur. Why is the tiger so steadfastly following the bear, while the bear hasn't sensed that it is being followed, or at least doesn't take avoiding action? The question nags at

Yellow-throated marten
This striking animal is mostly yellow, except for the dark brown legs, head, tail and hind quarters. The tail is almost as long as the body. On the ground it can jump two metres, and it readily leaps from tree to tree. Its main prey in winter is musk deer, and in summer animals up to the size of roe deer.

us, and we keep thinking up new explanations. It seems that the only way we shall get an answer will be to go on tracking the pair, however long it takes. I suggest this to my companion, but immediately reject it. All very well, but time is getting on, evening is approaching, and we still have to reach the cabin.

By the time we do reach it, the stars are out. In the dark I cut a hole in the ice, while he lights the stove and a paraffin lamp. After supper there is the usual checking of our observations before we turn in and sleep like bears.

In the morning we cook breakfast, tidy up in the hut, and set off along the path towards the foot of the pass over the Sikhote-Alin Range. The ascent makes itself felt long before it becomes really steep, while the path is still winding its way alongside a stream running down from the pass. It was not without reason that Arsenev named it *pereval Terpeniya* (Patience Pass).

'Here's a thought: why haven't I once seen tiger tracks on the pass? Have you?'

It seems my friend hasn't either, although he has been here several times in the last ten years.

'That could be important. No tiger spoor - that means ....'

'Hold on, don't jump to conclusions. We might still see some today.'

But we don't. Spoor of two wolves and a wolverine, and the successful hunt of three yellow-throated martens[8] for a musk deer, are clearly recorded in the snow on the east side of the pass. The half-obscured depressions which we at first take to be tiger spoor turn out, on careful examination, to be old bear footprints. Nor do we see the spoor we are hoping for among the spruce on the west slope beyond the pass.

'What would a tiger be doing up here? Hunting musk deer?'

'Maybe we can't expect signs of a tiger living up here, but what about animals crossing over from one river basin to another?'

'We haven't seen anything of the sort.'

'No, you're right.'

Well, we're now over half way, and our rucksacks are getting noticeably lighter. We continue along our set route, and next morning my colleague mentions home for the first time. We think of soon being back. The expectation of some exciting discovery reawakens once more, but the work of monitoring soon settles back into a routine.

On this side of the pass we have plenty to record. The taiga is so different from that on the east side. The valleys broaden out. Wide open snow-covered bogs with scattered trees glisten pure white. Instead of weaving between pines the path skirts around enormous white mounds, which are spruce trees looking as though they are fast asleep under their snow duvets. Places where there are wind-blown spruce with upturned root plates recall fairy-tale castles of the Snow Queen, or wood goblins' dwellings.[9] The river draining the western slopes is blocked in places with fallen timber and other debris, so we seek out clear areas of ice to look for spoor. We derive some comfort from knowing that, when this tributary joins the main river, we shall be traveling along the ice as easily as along a tarmac road.

However, we have to descend onto the ice earlier then we expected. From the bank we see on the snow-covered ice a tiger track. Like us, it is heading downstream, patiently working around all the obstructions. It went past several days ago. In many places the spoor has been flooded with water which then froze. Later, mink and sable ran along the same track. A mountain hare made us laugh out loud: for a long way it had run along the tiger track, jumping accurately from one tiger footprint to the next.

'It's chasing the tiger!'

The tiger consistently followed the ice on the river as far as its mouth. That is where our day's route ends - and then only one more to go. Part of it has been at night. We could have completed it sooner by taking the path, but the tiger was not aware of our requirements, or didn't understand them, and so has made us perform acrobatic feats when crossing piles of debris. Not for the first time we arrive at the hut long after dark.

It happened that the next - and last - day was the most interesting. We set off at a brisk pace, racing along on a thin layer of snow over ice. We noted the tiger track of the day before and, expecting nothing new, ski-ed alongside it talking out loud about what we'd seen. We came to the next bend in the river and suddenly stopped in our tracks and stared in amazement: there was a mass of tiger spoor - and quite fresh.

'How many, do you think?'

'We scared them off - our voices.'

We examined the ice over a sizeable stretch of river, bending down and staring at the snow.

'One, two, three. Three cubs and the tigress. Probably the same one we followed yesterday. Hey, look here!'

At the foot of a steep hill, where marsh grass *Calamagrostis* was covered in snow, I had stumbled on the place where the tiger family had spent the day. Or not just one day: it looked like a more permanent lair. There was no snow, the ground was trampled and melted, and nearby could be seen the trunk, almost white, of an ash stripped of its bark. Along the river were places where the cubs had lain down, turned somersaults, and fought each other. Here and there shone bright yellow tufts of fur.

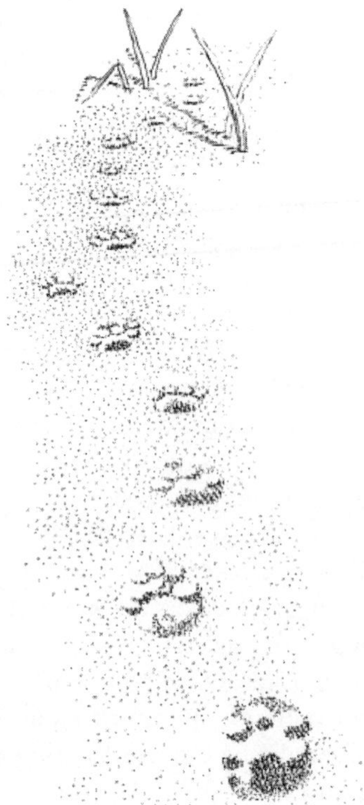

'Just look how much fur there is! Enough to make a shaving brush! This is amazing! What luck! We *would* have to charge along the river, shouting at the tops of our voices - while tigers scatter in all directions....'

'So, shall we follow them?'

'No, let's go in the opposite direction. We should first find out where they've come from, then where they've gone to.'

But it was no easy matter answering either question in detail. They had spent more than one day at the foot of the hill, and all the old spoor on the ice had been flooded by water, now frozen.

A vehicle was waiting for us in the evening, and in three hours we were home - back in the world of radios, electricity and clean sheets.

This account is the sort of thing that goes into our field notebooks, usually written in a laid-back style. The notes give a general impression of the animal life of the winter taiga. By sending monitors out simultaneously it is possible - from the size, freshness and direction of each set of spoor - to estimate the approximate size, and to some extent the sex- and age-structure, of the tiger population, and their distribution over the territory in one brief period - like a snapshot. Bringing together all available maps, details of sightings, and signs of tiger activity throughout the year, the zapovednik scientists can analyse them and estimate the state of the population in the zapovednik and surrounding districts. The results are written up in the annual Chronicle of Nature.[10]

By examining the Chronicle from 1935 onwards, we can get quite a clear idea of the turbulent half-century history of the recovery of the tiger population in the zapovednik - including episodes both encouraging and tragic. The story can be divided roughly into three periods.

1935-51: organisation and establishment of the zapovednik (interrupted by the war). The reserve's area was 1.8 million hectares. This vast area, in which all human activity was banned, helped the last 30-40 tigers in the southern part of the Soviet Far East to survive. Not only did they survive, they also spread out from the zapovednik.

1951-65: serious changes to the area and boundary of the zapovednik. For the first ten years the zapovednik covered 100,000 hectares (enough for only 3-4 tigers). On the rest of the original reserve intensive exploitation of natural resources began - mainly felling - with a corresponding serious decline in tigers. From 1963 to '65 the tigers in the zapovednik, by then increased to 310,000 hectares, were virtually absent.

Since 1966: period of reoccupation by the tigers of the protected and neighbouring areas. In these years the number of tigers in the zapovednik increased to 15-20 adults, with the number of cubs each year varying from 6 to 13.

What is particularly noteworthy is not so much the growth in the tiger population as the more frequent signs of their presence. In the zapovednik there are many places where tigers regularly wander and mark territory. Signs include paths with scrapes, trees blackened with scratch marks, and rocky outcrops with convenient niches where tigers make their lairs.

When you enter such a tiger area, you feel at every step the presence of tigers. Behind the next bush or tree, or on the rocks ahead you expect to see one of these magnificent cats.

Many foresters and scientists have not only seen tigers face to face, but have made a study of their routes and the regularity of their movements past guard posts and cabins, and along particular paths.

Of course all these observations are circumstantial, and sometimes we make mistakes. But the fact that, along a three-kilometre stretch of path by Silver Stream, you can at almost any time of year find 10-12 tiger scrapes, with nearby trees showing blackened bark from clawing, suggests that tigers have become permanent members of the fauna, along with the *izyubr*, wild pig, squirrel and sable.

Evgeni Smirnov's account shows very much how Kaplanov was working in the 1940s - except for the luxury of being picked up by vehicle at the end of the trek! The method has some use, but in the last half century better methods of studying the tiger have been developed, as described in the following paragraphs.

In the 1950s and 1960s a modest recovery in the number of Amur tigers in the Russian Far East took place as a result of making killing of tigers illegal and banning capture of cubs for 5 years. The recovery was monitored through further winter tracking surveys, especially by Kucherenko.[11] Since the second half of the 1950s the tigers have been counted at intervals not exceeding 7-11 years. These counts have used Abramov's methods of determining age and sex of tigers by the width of the pug marks in the snow, and through questionnaires completed by local hunters and other people living in the taiga.[12] The collection of data from local hunters was supplemented by regular observations throughout the winter and by recording tracks simultaneously in separate defined areas - as described by Evgeni Smirnov earlier in this chapter. In Soviet times considerable financial resources were devoted to these surveys, but it was not easy to continue them following the collapse of the USSR in 1991.

Despite the difficulties, in the winter of 1995-96 the most comprehensive census of the Amur tiger was carried out under the direction of Evgeni Matyushkin of Moscow University, who led an impressive team of some 650 trackers. This survey was funded by USAID and other international agencies. It was carried out with the co-operation of US scientists - in particular Dale Miquelle - who since 1992 have introduced advanced methods of radio telemetry and satellite mapping to the study of tigers and their habitat in the Sikhote-Alin.

Supplementing these extensive surveys, detailed studies of the ecology of the Amur tiger within more restricted territories have been carried out, in zapovedniks[13] and also in areas of the Sikhote-Alin range affected by human activities.[14] Information on the numbers of hoofed mammals, which form the basis of the tigers' diet, has been collected and analysed.[15] Regular observations of tiger paths, marking of trees and sites of tigers' prey have enabled specialists to analyse the changing structure of tiger populations.[16] Over a few decades, change can be discerned even in the tigers' attitude towards humans and their economic activities.[17]

In the development of really comprehensive plans for wildlife and programmes for the sustainable use of natural resources, the Hornocker Institute's activity in the Russian Far East set the standard for effective work. They have shown that a small, enthusiastic and efficient group can tackle immense conservation problems and achieve results by taking one step at a time. The Institute - based in Moscow (a town in Idaho, USA) and founded by the field biologist Maurice Hornocker - made its first contacts in the Russian Far East in 1989, during talks with the Soviet Academy of Sciences.[18] Although there were substantial problems when the USSR broke up, the institute committed itself to establishing an Amur Tiger Project, together with their Russian colleagues - notably Evgeni Smirnov and Dmitri Pikunov. The institute benefited from years of experience studying the mountain lion *Puma concolor* in the western USA. Field studies of the Amur tiger by the Institute scientists began in 1992.

Russian biologists had collected a great deal of valuable information by tracking tigers in the snow, but the Russian Government had not funded anything more sophisticated. The Hornocker Institute hoped to combine the Russian winter-tracking methods with radiotelemetry techniques. The tigers are trapped with snares - a noose foothold which does not cause injury - near trees that had been marked by them; they are then tranquillised and fitted with radio collars, which contain batteries and a small radio transmitter. The radio signals can be picked up by an observer on the ground holding an aerial, which gives the direction of the tiger from the observer. A different kind of collar uses the Global Positioning System and can record the latitude and longitude of the collared tiger several times each day; this may prove to be useful in the future.[19]

In February 1992 a young female tiger, called Olga, was the first to be captured and radio-collared by the Hornocker team working in the Sikhote-Alin Zapovednik. This tigress produced several litters and was still alive in 2003. By 1995 twelve tigers had been tranquillised and fitted with radio-collars in and around the reserve. This has provided the basis for a new study of Amur tiger ecology - including diet and hunting, home ranges, population genetics and dynamics, and the great cats' response to increasing human pressure.

Radio tracking has made possible improvements in the quality of knowledge about many aspects of tigers' lives. One reason is that they can be tracked at all times of the year, but also because they do not need to be disturbed so much. Take diet, for example: kills can be found more easily in summer, from which it has emerged that badgers[20] and bears - mostly hibernating during winter - are more important in tigers' diet than thought hitherto.

At the same time it is not necessary to disturb tigers on their kills as much as previously. Using traditional winter tracking, tigers were often disturbed while feeding, which led to low estimates of the amount of meat eaten from each kill, and also perhaps caused the tigers to kill more often. Improved information has been obtained for several other aspects of tiger ecology - such as disturbance to social structure by poaching, causes of mortality (mostly poaching in the study area), reproductive potential and resilience of the population, and habitat use - in particular tigers' reluctance to cross areas disturbed by humans, which effectively create barriers to their movements.[19]

Another approach has been applied to tigers more recently - scent discrimination. The method has been used previously by Russians, particularly Galina Salkina, but is being developed further by Linda Kerley, also at Lazovski Zapovednik. Trained dogs are used to identify which tiger has produced a sample of faeces (scat) or urine. First a set of identified samples are collected in the field - typically scats left by known tigers. If possible samples should be collected from all the tigers in the neighbourhood. Then, when an unidentified scat is found, it is brought back to the laboratory and placed in a dish, which the sniffer dog is trained to go to first. Having sniffed the sample, it goes around a series of dishes containing identified samples (from tiger no.1, tiger no.2, and so on), and sits down by the one with the same smell as the new sample. This identifies the tiger which produced the new sample. Eight individual tigers were identified in Lazovski Zapovednik over two seasons.[21]

This appears to be a powerful new approach to investigating tigers in the field. For one thing, it means that scats can be identified even in summer, when there are no tracks in snow, or any other means of identifying the tiger in question. It also means that provisional conclusions drawn from traditional tracking can be checked. For instance, it turns out that body size and size of pug mark are not closely correlated, so that earlier conclusions based on assuming that large spoor indicates an adult male may need to be revised. Some tigers just have big feet![22]

The object of the Hornocker Institute's Amur Tiger Project and Russia Program is to design and implement an effective conservation plan for the Amur tiger based on intelligent land-use planning and decision-making regarding natural resources in the Russian Far East. To conserve the tiger it is realised that the entire ecosystem has to be conserved and human needs integrated with conservation aims. In turn these aims must be understood and supported by local people and organisations.[23] Dale Miquelle has served as the Director of the Amur Tiger Project for several years as well as taking a lead in related programs, and John Goodrich is Field Director for the Project at Sikhote Alin Zapovednik.

Chapter 15

PROSPECTS FOR THE AMUR TIGER

By the 1930s the Amur tiger was brought to the brink of extinction by humans. In that decade the first tiger reserves were established in the Russian Far East, then in 1947 the Amur tiger became the first to receive state protection, and in recent years great efforts have been made to save it as a flagship animal and icon of conservation for the region.

These current efforts are reviewed in the following paragraphs - including the role of tigers in captivity, dealing with problem tigers, establishing and enlarging protected areas, political and economic development, tiger-human conflicts and the work of Russian and foreign non-governmental organisations (NGOs). Rather than describe each organisation and project when first mentioned, brief sketches and contact details of most NGOs are given in Appendix 3.

Captivity

There are today probably more Amur tigers in captivity than in the wild. Keeping them in unnatural conditions in captivity may seem to have little relevance to conservation, especially as many Amur tiger cubs were taken from the wild, which may have reduced the wild population. A fundamental principle of the Amur tiger captive breeding programme is that tigers should remain in the wild wherever possible, and fortunately the global captive population no longer requires further founder stock.

So, what is the justification for keeping them in captivity? In recent decades responsible zoos have made efforts to improve their standards of care and to turn their collections into centres for conservation, education and research.

Conservation. As Amur tigers are critically endangered in the wild, managed breeding programmes aim to perpetuate a healthy captive population, physically and behaviourally as near as possible to its original wild form and with comparable genetic variation, as a hedge against extinction in the wild.

As a result of careful management Amur tigers form one of the most numerous and genetically healthy of any captive animal.[1] At the end of 1996 about 500 registered Amur tigers were in captivity worldwide, excluding other unregistered animals.[2]

Education. Captive Amur tigers are charismatic animals and powerful icons of nature conservation, playing an important educational function as ambassadors for their kind. So exhibiting Amur tigers in good conditions can stimulate public concern about the plight of the tiger in the wild and raise funds for tiger conservation.

Research. Observations on captive specimens have helped understanding of tigers in the wild.[3] For instance the European Endangered Species Programme has been able to assist field workers from the Amur Tiger Project in Russia by collecting footprint measurements from tigers of known sex and age for the 1995-96 survey in the Russian Far East, and by collecting tiger faecal samples from animals of known identity to help test sniffer dogs. Hopefully the two programmes of wild and captive protection can reinforce one another, to the tigers' advantage. Captive tigers have also been used to develop anaesthetics for use in fitting radio-tracking collars and in moving tigers to new sites.

Can captive-bred Amur tigers be successfully released into the wild? There *have* been individual examples of big cats learning to hunt as adults after being raised by humans. However, although much initial hunting behaviour is instinctive, young tigers learn some of their hunting skills from their mothers and others by trial and error. It would be a challenge for captive Amur tigers to cope with the rigours of life in the wild.

Finding and killing prey require experience, strength and skill, and captive-bred animals might face competition from other tigers. Released tigers - whether captive-bred or wild animals moved from another area - will disrupt any existing tiger social structure and are likely to be killed by resident tigers if the territory is already occupied.[4]

Tigers that have grown up in captivity and so have lost their fear of humans are unlikely to be suitable for reintroduction into the wild. Each winter there are cases of tigers entering villages in the Russian Far East and attacking domestic animals, and this would certainly happen if released tigers are already used to humans.[5]

For problem tigers over a year old, captivity is not a satisfactory answer: mature wild-caught tigers do not respond well to captivity, as may be seen in Nepal where formerly man-eating tigers have been brought into Kathmandu Zoo. Tigers are expensive to keep in zoos, and breeding programmes already have a sufficient number of founder animals.

Nonetheless in extreme cases it may be necessary to bring Amur tigers into captivity. In January 2001 two 4-5 month old tiger cubs, one with badly infected facial injuries, were found abandoned under a village building. Both were seriously underweight. The mother of the cubs had been driven away by villagers after it had killed a village dog. As she could not be found and the cubs were too young to survive in the wild on their own they were treated, fed and sent to Moscow Zoo. There the injured cub's damaged jaw was operated on, and three molar teeth were taken out. Both tigers made a full recovery. By the summer of 2001 they were thriving, and were admitted to the European Endangered Species Programme, which does its best to find suitable homes for such animals.[6]

Problem tigers

There are a number of options in dealing with problem tigers, including orphaned cubs.[7] The simplest is just to take no action. The problem tigers may be youngsters wandering about looking for a territory, and an attack on domestic stock is not always followed by further attacks. Even orphaned or abandoned cubs may sometimes be left in the wild, since they can survive by preying on smaller animals such as raccoon dogs, hares or badgers. Cubs can sometimes be returned to their mother, though this is not always successful. Marauding tigers may be driven away by rockets or other means; this has sometimes been successful.

Another option is to provide extra food until young tigers become competent hunters. In 1998-99 three orphaned cubs were given additional food by the Amur Tiger Project team for three months, after which they appeared to be self sufficient and dispersed to live independent lives.

There have been several documented cases of problem tigers in the Russian Far East being moved to new areas. In 1975 a large male tiger, which had killed several dogs near a town, was immobilised, sent by boat to Vladivostok and from there flown by helicopter to the Iman valley. It was tracked for some time by tiger expert Nikolayev, but the trail was subsequently lost. In December 2000 the Tiger Response Team[8] captured a two-year-old male tiger, which had been preying on dogs in a village, and released it 150 km to the east. However, it continued to prey on dogs and was shot by the caretaker of some dachas. In this case the problem was not solved, but simply transferred to another locality.

In July 2001, after several months of secluded captivity and rehabilitation, two wild-caught female tigers were released into the wild. The first, named Troya, was discovered in June 2000, when less than a year old, entangled in a fence near a village on the Amur River. She was taken to a rehabilitation centre. The second, slightly older female called Pakhomovna, was found with an injured paw and captured in a kennel while feeding on a dog it had killed in another village. After being rehabilitated in natural wooded enclosures, completely secluded from people, the two tigers were fitted with radio collars and released in July 2001, some 10 km apart in an area just south of Sikhote-Alin Zapovednik. Troya was about two and Pakhomovna nearly three years old. Pakhomovna moved into the protected area of the zapovednik, established a territory of her own and was monitored by scientists after her release. Sadly she was lost to poachers after one year. Troya moved away in a different direction, contact was lost after a typhoon, and it was assumed that she also had fallen victim to poachers.[9]

In July 2002 a forestry worker in southern Khabarovsk Region was injured by a one-year-old male tiger; it was trapped, found to have no injuries, and released into the wild after a few days of rehabilitation.[10]

Although - mainly because of poaching - this programme has had mixed results translocation remains a feasible option in areas of the Russian Far East where there are not too many people or other tigers.

As a last resort it may be necessary to destroy some problem animals. In the winter of 2000-01 three tigers were killed by the Tiger Response Team in the Russian Far East. One was a decrepit old female with severe dental disease, the second was an old male repeat-offender with broken canine teeth, and the third attacked a Tiger Response Team member, having already mauled a forestry worker.

Protected areas

The system of protected areas is crucial to the tigers' survival. The average size of a mature female's home range is about 450 sq km, and to protect the genetic diversity of the population in the short term it has been suggested that 50 territories be protected. Ideally 300 territories or 135,000 sq km would be needed; however, this would require 70% of the whole of Primore and is hardly feasible! Even with the enlargement of protected areas to 25,600 sq km, this would still leave 250 out of 300 adult female ranges unprotected.[11] So the kind of land use in each territory will determine the well being of the population in the future. In 2002 the Governor of Primore announced plans for a threefold extension of protected areas in the territory, though this will take some years to complete.[12]

Following earlier proposals by the Russian regional authorities, Russian and US scientists have drawn up a master habitat-protection plan aimed at saving what remains of the Amur tigers' hard pressed domain. The plan calls for an inviolate core, in the form of a network of protected areas linked by corridors to allow safe dispersal of young tigers. Around the core, unprotected forest will be carefully managed to ensure that logging, mining and construction do the least damage to the habitats of the great cats and their prey.[13]

Across the frontier in Northeast China some good tiger habitat remains and reserves have been set up, since tigers still cross over from Russia. A crucial task is the restoration of populations of prey animals, particularly on the Chinese side of the border.[14]

Politics and economics

In the early 1990s Amur tigers faced serious threats as hard economic conditions accompanying the collapse of the Soviet Union forced many local people to rely on forest products and hunting for meat. Unregulated logging and mining for minerals also threatened the tigers' habitat. With the growth of organised criminal gangs and weakening of law and order the tigers were decimated in a great wave of poaching between 1992 and 1994.

However, after great efforts by the Russian authorities supported by international conservation bodies between 1995 and 1999, the situation seemed to have improved. In 1995, following intensive international lobbying, the Russian Prime Minister Viktor Chernomyrdin decreed a national conservation strategy for the tiger. Patrolling was strengthened, the frontier with China was better regulated, and poaching was reduced.

The painful transition from a Communist state-controlled command economy to a potentially democratic political system and market economy presents both challenges and opportunities regarding the conservation of Russia's natural resources. A combination of severe economic depression, inflation and the breakdown of legal controls reduced the ability of central and regional government adequately to fund nature protection, including the zapovedniks.

President Vladimir Putin signed a decree on 17 May 2000 abolishing the State Committee for Environmental Protection and the Federal Forestry Service. Their powers were transferred to the new Ministry of Natural Resources.[15] One immediate effect of the changes has been to cut government funding for Inspection Tiger,[16] reducing the number of paid members in each team from 4 to 3, thus limiting their effectiveness. Fortunately in 2001 regional and overseas NGOs stepped in with funding to make up the team numbers again by recruiting and training university students for this demanding task. The tiger protection organisations in the Russian Far East are now working closely with the new Ministry of Natural Resources.

In August 2002 President Putin recognised the importance of nature conservation by visiting the Marine Zapovednik near Vladivostok. Despite well-meaning pronouncements and gestures supporting wildlife by federal and regional governments, both internal and external pressures for greater economic exploitation of natural resources are bound to intensify, with consequent threats to natural habitats. Even so, the situation is not just one of 'mafia' crime, the draining of capital abroad, asset stripping, corruption and poverty for the majority of the people. There are some signs, albeit limited, of economic improvement, with significant growth in the Russian economy in 1998-2001. While on the one hand economic growth may lead to environmental damage, successful economies also have the means to pay for nature protection. Already WWF-RFE has secured financial sponsorship for tiger protection from members of the business community, including a leading brewing company.

Tiger-human conflicts

The tiger's future is very much tied up with its relationship to the human population. If this can be successfully managed, there is some hope. Tiger-human conflict varies markedly with food availability, and natural conditions for the tigers' prey can fluctuate greatly from year to year. Snowy winters and seasons of poor acorn and pine-kernel crops may be disastrous for hoofed animals, and therefore also for tigers.

The winters of 1867 and 1914, for example, were critical for hoofed animals and tigers, despite the fact that most of the forests were then much like the few remaining in the present nature reserves. In 1867 - when the number of Amur tigers in Russia was perhaps 1000 - there was a terrible famine.[17] In that year 21 people were killed by tigers in one district alone. In the winter of 1914-15 virtually all the wild pigs perished in heavy snow, and red deer numbers plummeted. Again tigers went into villages, killed people and were shot.[18] Tiger-human conflict has been repeated during severe winters ever since.

In the southern Russian Far East the winter of 2000-01 was such a case, with deep snow and temperatures falling below -25°C for long periods. With conditions of frozen crunchy snow the tigers were not able to approach their prey silently. In Khabarovsk Region the exceptionally harsh winter, made worse by deforestation and over-hunting, reduced the number of prey. This resulted in six incursions into one village by tigers and the killing of three dogs. Yuri Dunishenko, a leading tiger expert in Khabarovsk Region, estimated that 80% of tiger cubs in the area had perished in the winter cold.[19]

With greater interaction between tigers and humans in the Russian Far East, there were several cases of Amur tigers killing people in 1997 and '98, though these were usually of poachers who were hunting the tigers or their prey. In one incident two hunters were killed and eaten by a tiger. From November 2000 to January 2001, Anatoly Medetsky reported 8 attacks in one district of Primore by tigers on cattle, sheep, dogs, and a forestry worker. Rangers were forced to shoot two emaciated tigers and two more were given radio collars for study purposes.[20] One very thin tiger was discovered in the basement of a villager's house in the Terney area and the full grown animal had to be moved quickly to a veterinary centre in the back of a car before the anaesthetic wore off!

Thus tiger-human conflict is a continuing problem in the region and a great concern for local conservationists. To cope with it the Russian authorities set up a special Tiger Response Team and appealed for funds to provide special crates for moving animals at risk. Five British and seven European zoos responded: the crates were built at Moscow Zoo in 2001 and have been in use.

The winter months of January-March 2002 in the Russian Far East were severe, with snow a metre deep. There were serious losses of roe and young sika deer in Khasanski District; being approachable by people to within 2 or 3 metres, they were easy prey for human hunters. The large number of deer that perished through starvation may have initially provided some meat for tigers and leopards, but in the following season the effect was to reduce the cats' food supply. Twelve tons of hay and cut browse were provided as supplementary food for the wild deer and paths were made through the snow by the nature-protection authorities and NGOs such as the Wildlife Conservation Society and WWF-RFE. The State Hunting Department agreed to a hunting ban for at least the summer of 2002.[21] There have been severe winters in the past and most adult tigers are strong enough to survive them, providing that human hunting pressure is not too severe.

Russian and foreign NGOs

If the Amur tiger and its diverse ecosystem are worth conserving, then perhaps the rest of the world should help pay the cost. In the years following 1993 there was considerable growth in international support for the tiger.

In 1998-2000 the World Wide Fund for Nature, especially from Germany, contributed more than half of over £3 million allocated by NGOs to Amur-tiger conservation projects.

21st Century Tiger is another organisation playing an important strategic role in channelling funds to the Russian Far East to help the tiger; in 1998, emergency fire-fighting equipment was purchased for Sikhote-Alin Zapovednik during devastating forest fires resulting from unusually dry weather.

Anti-poaching patrols and customs control, especially at the frontiers and ports, are vital elements in combating the illegal trade in tiger parts. Anti-poaching efforts have been supported by several western organisations, including the Save the Tiger Fund, United States Aid for International Development (USAID), and WildAid;[22] 21st Century Tiger has funded a radio communication system to link the anti-poaching teams over great distances.[23] But poaching for tiger pelts and body parts and their traffic continues. From 1994 to '99, 13 out of 21 radio-collared tigers were lost to poachers, and even more tigers without radio collars were killed.[24] However, thus far there have been few successful prosecutions of tiger poachers in the courts.

In 2001 21st Century Tiger supported a GIS mapping project that will provide a detailed map of tiger habitats and human impacts in order to identify key areas still unprotected, zones vulnerable to habitat fragmentation, and areas where reducing human impacts could substantially help tiger populations. The project also provides equipment and training for Russian scientists at various institutes doing the work. Linked to this study is a comprehensive survey of hoofed-animal habitats. 21st Century Tiger invested £61,000 on these two projects running throughout 2001.[25]

Saving the Amur tiger must depend ultimately on the Russian people, especially those living within the regions where the tiger exists. This requires the involvement of the public as well as fundraising within the country and commitment from federal and provincial governments. To help with this an Anglo-Russian charity AMUR was successfully launched in May 2001 at the British embassy in Moscow.

In 2002 most of the agencies already working closely together in Amur-tiger and leopard protection formally agreed to form the Amur Leopard and Tiger Alliance (ALTA). This manages about half of the tiger and leopard projects in the Russian Far East, and works closely with WWF and the Russian nature authorities and scientific institutes.

Conservation in practice

In addition to essential work in maintaining and if possible expanding the protected areas, tiger conservation is taking a number of forms.

Hunting for meat and for furs is very much part of the way of life of Far Easterners and much of the region is leased for hunting. Controlled hunting is not necessarily bad for the tiger as the natural habitat is preserved and resident hunters soon become aware of poachers; legitimate hunters can in effect become 'policemen'. Efforts are being made to set hunting quotas for the tigers' prey through cooperation between conservation and hunting organisations.

Attempts are also being made to control illegal logging and organise more effective fire-fighting. People suffering from depredations by tigers and leopards have been compensated, and farmers have been paid to retain tigers and leopards on or near their land. Deer farms have been leased or purchased, and managed in a manner friendly to tigers and leopards. Orphaned or injured young tigers have been rehabilitated and translocated. These measures are supported by detailed **research** on the tigers' ecology by the Amur Tiger Project supported by the Wildlife Conservation Society.

Education programmes on nature in the region are being extended. Annual Tiger Day festivals in Vladivostok and other centres have become popular events focusing on saving the Amur tiger and other natural treasures of the Russian Far East. A Russian NGO, the Phoenix Fund, coordinates conservation activities in the region and develops contacts with the mass media. Particularly important in this regard is the Vladivostok journal *Zov taigi* (Roar of the Taiga), a leading voice of the local conservation movement.

Eco-tourism in this beautiful region remains a possibility for the future, though tourist infrastructure is limited. Some reserves, for instance Ussuriyski Zapovednik,[26] do encourage visitors, including overseas students, to appreciate the beautiful forests and wildlife of Primore.

Working with the **local human population** may not have been given sufficient weight in the past. In December 2002 Anatoly Lebedev of the Primore Nature Protection Society was reported as saying that the number of Amur tigers in the Russian Far East had declined from 400-450 in 1998 to no more than 190 in 2002.[27] Despite more than $450,000 provided by international agencies in the first two years from 1998, tigers were still being killed, mostly by local residents. Lebedev argued that the money would have been better spent, at least in part, on improving the social and economic conditions of the people in forest settlements in the region, rather than just chasing poachers.

The following day these criticisms were strongly rejected by Yuri Darman of WWF-RFE. He pointed out that winter monitoring of the wild tiger population in the Russian Far East in February 2002 from 16 survey stations showed numbers to be stable at around 400-450, and that the number of tiger pelts recovered in 2002 by Inspection Tiger, customs control and the police was about average, at seven for the year.

The future

Since human pressures will probably continue increasing, the survival of the Amur tiger must remain problematic. Especially worrying is the continuing poaching of tigers for their pelts and body parts and the reduction in the number of prey through over-hunting. Logging, clearing land for agriculture, burning for pasture, and the spread of settlement along river valleys threaten the integrity of the great continuous forest of the Sikhote-Alin.

Can people and tigers coexist? Or will the Amur tiger share the fate of the other tiger from the former Soviet Union - the extinct Caspian tiger? Much may depend on how far economic development can be reconciled with the needs of the great striped cat.

The survival of such a high-profile animal as the Amur tiger is not just an issue of wildlife protection. Although the majority of Russian Far East citizens regard the tiger positively, many of the threats to wildlife in the region result from the poverty of local people. Local people need to put food on their tables and, so long as incomes are depressed and unemployment remains high, natural resources will continue to be taken from the forest. It may well be argued that while the focus has been on tiger protection too little money has been spent by international agencies on improving the lives of people in remote settlements. Sustainable development - through projects such as woodworking, bee keeping and eco-tourism, especially in the villages - are likely to improve people's welfare, at the same time as protecting the wildlife and forests. By involving local people to a greater extent in conservation programmes and improving their quality of life through sustainable economic development, there will be a greater chance of reconciling people with the tiger and saving the rich diversity of animal and plant life in the Ussuri taiga.

I vividly recall my first sight in 1992 of the vast primeval forest of the upper Bikin valley, the 'forest sea', like undulating waves, stretching as far as the eye could reach eastwards to the Sikhote-Alin range. At dusk we heard, above the other sounds of the forest, the roar of Amba, the great Amur tiger. The hope must be that these treasures of nature will be saved for future generations to appreciate.

Appendix 1: Species Mentioned in the Book

This list gives only a selection of the various names that have been used for the animals mentioned, and some Latin names have variant spellings, eg *man(d)shuricus, t(h)ibetanus*. The scientific name is followed by English and Russian vernaculars, with sometimes further variants after a colon. Not all the species are native to or even found in the Russian Far East. Further information can be found via the index.

MAMMALS - predators/omnivores

Canis lupus, Grey Wolf, *volk*

Cuon alpinus, Asiatic Wild Dog or Dhole, *krasny volk*

Desmana moschata, Desman, *vykhukhol'.*

Enhydra lutris, Sea Otter, *kalan.*

Prionailurus bengalensis euptilura, Far East or Amur Forest Cat or Leopard Cat, *dal'nevostochny kot.*

Lynx lynx, Far-Eastern Lynx, *rys'*; also *Felis lynx stroganovi.*

Gulo gulo, Wolverine, *rosomakha.*

Lutra lutra, Otter, *vydra.*

Martes flavigula, Yellow-throated or Indian Marten, *kharza* or *kunitsa.*

Martes zibellina, Sable, *sobol'.*

Meles meles amurensis, Badger, *barsuk.*

Mustela erminea, Common or European Stoat, Ermine, *gornostay.*

Mustela lutreola, European Mink, *obyknovennaya norka.*

Mustela nivalis, Least Weasel, *laska.*

Mustela sibirica, Siberian Weasel or Kolinski, *kolonok*; also *Kolonocus sibiricus.*

Mustela vison, American Mink, *amerikanskaya norka.*

Nyctereutes procyonoides, Raccoon-dog, *yenotovidnaya sobaka.*

Panthera pardus orientalis, Amur or Far-Eastern Leopard, *leopard.*

Panthera tigris, Amur or Ussuri Tiger, *amurski/ussuriski tigr*; subspecies name '*altaica*' no longer valid.

Ursus arctos, Brown Bear, *bury medved'.*

Ursus thibetanus, Asiatic, Asian or Himalayan Black Bear, *gimalayski* or *belogrudy medved'.*

Vulpes vulpes, Red Fox, *lisitsa.*

MAMMALS - herbivores

Alces alces cameloides, Manchurian Elk or Moose, *los'.*

Axis axis, Chital, Spotted Deer.

Axis porcinus, Hog Deer.

Bos gaurus, Gaur.

Bos grunniens, Yak.

Bos javanicus, Banteng.

Bos taurus, Domestic Cattle.

Capreolus pygargus, Siberian Roe Deer, *sibirskaya kosulya.*

Castor canadensis, North American Beaver, *bobr*.
Cervus canadensis xanthopygus, Manchurian Wapiti, *izyubr'* or *izyubr*.
Cervus duvauceli, Barasingha, Swamp Deer.
Cervus elephas bactrianus, Bactrian Red Deer.
Cervus nippon mantchuricus, Sika Deer, *pyatnisty olen'*.
Cervus timorensis, Rusa or Timor Deer, Sunda Sambar.
Cervus unicolor, Sambar.
Elaphodus cephalophus, Tufted Deer.
Equus caballus, Horse.
Lepus brachyurus, Japanese Hare, *kustarnikovy zayats*.
Lepus timidus, Mountain Hare, *zayats-belyak*.
Moschus moschiferus parvipes, Far-Eastern Musk Deer, *sibirskaya kabarga*.
Muntiacus spp, Muntjac.
Nemorhaedus caudatus raddeanus, Amur or Korean Goral, *goral*; also *N.crispus*.
Ondatra zibethica, Muskrat, *ondatra*.
Rangifer tarandus, Reindeer, *severny olen'*.
Sciurus vulgaris, Red Squirrel, *obyknovennaya belka*.
Sus scrofa, Wild Pig or Wild Boar, *kaban*.

BIRDS

Bubo bubo, Northern Eagle Owl, *filin*.
Glaucidium passerinum, Pygmy Owl, *sych-vorobey*.

PLANTS - trees & shrubs

Abies nephrolepis, East Siberian or Khingan Fir, *belokoraya pikhta*.
Betula dahurica, Dahurian birch, *chernaya bereza*.
Fraxinus mandshurica, Manchurian Ash, *man'chzhurskaya yasen'*.
Juglans mandshurica, Manchurian Walnut, *orekh*.
Larix gmelinii, Dahurian Larch, *listvennitsa Gmelina*.
Phellodendron amurense, Amur Cork Tree, *amurskiy barkhat*.
Picea ajanensis, Yeddo Spruce, *ayanskaya yel'*; also *P.jezoensis*.
Pinus koraiensis, Red or Korean Pine, *koreyskiy kedr*.
Pinus pumila, Siberian Dwarf Stone Pine, *stlanik*.
Populus tremula, Aspen, *ocina*.
Quercus mongolica, Mongolian Oak, *mongol'ski dub*.

PLANTS - herbs

Calamagrostis sp., Small-reed, *veynik*.
Panax ginseng, Ginseng, *zhen'shen'*.
Pteridium aquilinum, Bracken, *orlyak*.
Ranunculus (Batrachium) sp., Water Crowfoot, *vodyanoy lyutik*.

Appendix 2: The Different Kinds of Tiger

Abbreviations: RFE - Russian Far East; TCM - traditional Chinese medicine. Note: the Latin names given here are those long used for the eight 'subspecies' of tiger. Five may however be invalid, if only three subspecies are now recognised as a result of recent research (*tigris*, *virgata* and *sondaica/sumatrae* (Kitchener & Dugmore 2000)).

Bali Tiger *Panthera tigris balica* (Schwarz 1912).
Distribution: Bali, Indonesia.
Habitat: moist tropical evergreen forest, tall grasslands, mostly in the west of the island.
Morphology: described as smallest form with comparatively dark, rich orange colouring with widely spaced stripes in some specimens. But tigers show great individual variation in colour and stripe pattern. Adult male 100kg. Total length 240-250 cm (Mazak 1981).
Main prey: like all tigers it preyed on large ungulates, especially deer and wild pigs, Rusa or Timor deer, banteng.
Status and threats: extinct by 1940s. Because of the size (5500 sq km) of Bali, there may have been only 125 or so adult tigers there in the 19th century. Played an important part in the rich Hindu culture of Bali. Isolation on a relatively small island with expanding human population, loss of habitat for agriculture, and hunting pressure led to extinction.

Javan Tiger *Panthera tigris sondaica* (Temminck 1844).
Distribution: Java.
Habitat: tropical moist evergreen forests.
Morphology: dark orange in colour and heavily striped. Similar in morphology to Bali tiger but perhaps slightly larger. Descriptions based on very few specimens. Adult male 100-140 kg. Total length c 248 cm.
Main prey: Rusa deer, banteng, wild boar.
Status and threats: extinct in 1980s. Abundant in 19th century. Survived only in a few remote forests by the 1940s; could not survive in a few restricted reserves. With rapidly growing human population, expanding rice cultivation resulted in massive deforestation in 20th century. Intensive hunting, including poisoning ,for skins and TCM.

Caspian (or Turan) Tiger *Panthera tigris virgata* (Illiger 1815).
Distribution: previously ranged from eastern Turkey to Lop Nor in Xinjiang, China. Especially northern Iran, former Russian and Chinese Central Asia and Transcaucasia (especially in Lenkoran mountains, Azerbaijan).
Habitat: along river valleys and around lakes especially in reed beds, tamarisk and other tugay vegetation.
Morphology: long, soft, plush winter coat, long belly and cheek fur. More numerous closer striping, more reddish background colouring than Amur tiger; however, as with other tigers, considerable individual and local variation in colour and stripe pattern. Adult male 170-240 kg. Total length 270-290 cm.
Main prey: wild pig, Bactrian red deer.
Status and threats: extinct by 1950s in most of its range but a few may have survived to 1970s in northern Iran. Restricted band-like distribution in river

valleys, around lakes and oases. Riparian tugay habitat vulnerable to human expansion, especially in regions which had been cleared of reeds for cotton cultivation. Celebrated in Islamic Sufi art.

South China (or Chinese) Tiger *Panthera tigris amoyensis* (Hilzheimer 1905).
Distribution: south-east-central China. Previously in Hunan, Fujian, Hubei, Jiangxi and Zhejiang provinces.
Habitat: sub tropical and temperate montane forest.
Morphology: comparatively small, orange-ochre background colour, widely spaced stripes. 130-175 kg. Total length 230-280 cm. The Chinese tiger is smaller, less massively built and darker in colour than the Amur tiger. With its shorter braincase and more forward facing eye sockets, Herrington (1987) has suggested that it may be the prototype or stem tiger population, though Kitchener (1999) has questioned this idea.
Main prey: wild pig, muntjac, sambar, tufted and roe deer, serow.
Status and threats: facing extinction, perhaps 20-30 left; no evidence of their existence found in extensive 2001-02 survey. Reduced from 3000-4000 in 1940s by hunting in 1950s and '60s as a pest and for skins and products for TCM. The tiger has played an important part in Chinese culture. Seventy-five or so exist in Chinese zoos, but are inbred and show poor fertility; this and inadequate natural prey in the wild make prospects for reintroduction to the wild poor.

Sumatran Tiger *Panthera tigris sumatrae* Pocock 1929.
Distribution: Sumatra.
Habitat: tropical forests, both primary and secondary, *alang-alang* tall grassland, freshwater swamps.
Morphology: relatively small tiger, shorter in the leg than northern tigers, in many cases dark orange with close black stripes, long spiky cheek fur. 100-140 kg. Total length 220-255 cm.
Main prey: wild pig, sambar.
Status and threats: about 400-450 remaining. Much reduced from earlier numbers (thousands in early 1900s, c.1000 in 1960s) by poaching, destruction of habitat through deforestation and spread of settlement. Survives in several large national parks with intensive protective measures.

Indo-Chinese Tiger *Panthera tigris corbetti* Mazak 1968; named after the famous tiger hunter Jim Corbett, author of *Maneaters of Kumaon* (1944) and *Man-eaters of India* (1957). It is questionable whether this tiger can be distinguished taxonomically from the Indian tiger.
Distribution: mainland South East Asia. Eastern Myanmar, Thailand, Malayasia, Cambodia, Laos, Vietnam, southern Yunnan province of China.
Habitat: tropical moist and dry deciduous forests, montane forests.
Morphology: mostly smaller than Indian tiger, but similar orange red background colouring with black stripes, white underside. 150-195 kg. Total length 255-285 cm.
Main prey: Sambar deer, muntjac, wild pig.
Status and threats: estimated at 1100-1700 but numbers not really known. Everywhere declining through poaching, loss of prey and habitat. May be reasonably well protected in larger national parks of Thailand and Malayasia.

Indian (or Bengal, Sunderbans) Tiger *Panthera tigris tigris* (Linnaeus 1758).
Distribution: India, Nepal, Bangladesh, Bhutan, western Myanmar, southern Tibet.
Habitat: tropical moist, evergreen and dry deciduous forests, alluvial grasslands, mangroves.
Morphology: one of the largest tigers. Pale tawny-yellow to orange-red background colouring. Usually finely and fully marked with black stripes (but this varies widely). White underside. 180-258 kg. Total length 270-310 cm.
Main prey: Sambar, chital, barasingha and hog deer, wild pig, gaur; yak, cattle and horse in Tibet.
Status and threats: most numerous tiger population, but much reduced in number and fragmented since early 20th century. Estimates: 2500-3500 in some 60 (often isolated) protected areas or their vicinity. Losses continue through poaching for hides and for TCM, reduction of habitat (pressure of expanding human population) and reduction of prey. Important in Hindu art and religion.

Amur (or Siberian, Northeast, Manchurian) Tiger *Panthera tigris altaica* (Temminck 1844). Type specimen from Korea; named *altaica* after mountains in Korea, not the Altai Mountains of Mongolia (Mazak 1967).
Distribution: Russian Far East (Primore and southern Khabarov Krai), mostly in the Sikhote Alin range. A few in adjacent areas of Northeast China and Korea.
Habitat: mixed temperate broadleaved-coniferous forests.
Morphology: rivals or exceeds Indian as largest tiger. Massively built. Usually ochre-yellow background colour with fewer more widely separated black stripes than in Indian form. More white on underside. Long, thick, silky winter coat. Lays on fat in winter as insulation. 180-306 kg. Total length 270-330 cm.
Main prey: wild pig, sika, roe and musk deer; occasionally elk (in the north), gorals (in the south), black and brown bears, *izyubr* .
Status and threats: probably 600-800 in RFE in mid 19th century (maybe more in NE China at that time). Reduced to c 30 in RFE in 1940s, with c.12 in China along the Russian border. First tiger population to be legally protected (1947). Perhaps only tiger population to show an increase in 1950s and 1960s (increase continued 1970s and '80s). Population in 2001 small but probably stable (430-470 in 1996). Great efforts are being made to protect it in the face of (a) poaching for pelts and for bones especially for TCM, (b) the serious decline of prey animals and (c) destruction of habitat. Totem animal for Udege tribe.

Appendix 3: Projects, NGOs and Other Organisations

AMUR

An Anglo-Russian, British-registered charity founded in 2001 to preserve Amur tigers and leopards through fund raising, exhibitions and publicity in the mass media. It has growing support from the Russian business community and sponsorship from a number of British companies. It works in partnership with the Russian authorities - including federal and regional government - and also scientists and other conservation organizations.
www.amur.org.uk

Amur Leopard and Tiger Alliance (ALTA)

An umbrella organisation for Phoenix, AMUR, Wildlife Conservation Society (WCS), WildAid, Tigris Foundation, 21st Century Tiger and Zoological Society of London.

Amur Tiger Programme

Research project on the Amur tiger and other species based in the Sikhote Alin Zapovednik; it was established in 1990 by the Hornocker Wildlife Institute (www.hwi.org), which is now merged with the Wildlife Conservation Society.

Cat Specialist Group

A component of the Species Survival Commission (SSC), and one of 6 volunteer commissions of IUCN, the World Conservation Union, founded in 1948. SSC's mission is to conserve biodiversity by developing and executing programmes to save, restore and manage species and their habitats. SSC members provide technical and scientific advice for conservation projects throughout the world.
Thunstrasse 31, CH-3074 Muri b. Bern, Switzerland
www.catsg.org

Phoenix Fund

A Russian non-profit NGO, founded in 1998. It supports and co-ordinates biodiversity conservation projects in the Russian Far East, including support for anti-poaching teams, educational projects, fire fighting, customs control and compensation for predator damage. Funded by Tigris Foundation, Save the Tiger Fund, 21st Century Tiger and International Fund for Animal Welfare.
Room 409, Petra Velikogo Str 2, Vladivostok 690091, Russia
www.phoenix.vl.ru

Russian Conservation News

A quarterly magazine covering all aspects of nature conservation in the former Soviet Union.
c/o Margaret Williams, POBox 57277, Washington DC 20037-7277
e-m rcn@igc.org (USA) or rcn@online.ru (Moscow).
www.russianconservation.org

Russian Far East Non Timber Forest Products Processors Association.
Founded in 1996, it organises and encourages the production and marketing of non-timber forest products by villagers in a sustainable manner.
71 Volochaevskaya St, Khabarovsk 680030, Russia
www.ntfp.ru

Save the Tiger Fund
A joint effort of the National Fish & Wildlife Foundation and Exxonmobil, Washington, USA.

Tiger Information Centre, run by Minnesota Zoo, USA
An information centre providing news and research reports on tigers worldwide
www.5tigers.org

Tigris Foundation
A Netherlands NGO, founded in 1996, dedicated to the Amur tiger and leopard, with the aims of developing new conservation projects and strategies in the Russian Far East and raising funds for their implementation, especially in anti-poaching activities.
www.tigrisfoundation.nl

Wildlife Conservation Society
Founded in 1895 and formerly known as the New York Zoological Society, it has amalgamated with the Hornocker Institute and now organises field research on the Amur tiger based at the Sikhote Alin Zapovednik. The Amur Tiger Programme/Project is one of many organised by the WCS worldwide.
Wildlife Conservation Society, Bronx Zoo, New York. 10460-9973, USA
www.wcs.org

World Wide Fund for Nature WWF.
The largest conservation NGO in the world, with an active organization based in Vladivostok:WWF-RFE (www.wwf.ru). It has provided more than half of the funding for tiger protection in the Russian Far East, notably for anti-poaching teams and educational work.
WWF UK, Panda House, Weyside Park, Godalming, Surrey GU 7 1XR, UK
www.wwf.org.uk

21st Century Tiger
A funding coalition, founded in 1997 and based at the Zoological Society of London; includes Global Tiger Patrol; supported by Esso Corporation. It works together with the international conservation and scientific community to save the tiger, and raises the profile of tigers through public awareness and educational programmes; especially involved in the Russian Far East and Sumatra.
21st Century Tiger, c/o London Zoo, Regent's Park, London NW1 4RY, UK

Appendix 4: Where to see Amur tigers

The following is a list of zoos in the United Kingdom and Ireland exhibiting *registered* Amur tigers as of July 1 2003. The number of tigers in each collection sometimes changes.

Banham Zoo, The Grove, Banham, Norfolk NR16 2HE.
9 miles northeast of Diss, signposted off A11 and A140. 30 min. driving time from Norwich.
Tel. 01953 887771. Web www.banhamzoo.co.uk.
Amur tigers two - 2 males.

Blackpool Zoo Park, Blackpool, Lancashire FY3 8PP.
3 miles east of seafront on East Park Drive, A587. Near Stanley Park.
Tel. 01253 830830. Web www.blackpoolzoo.org.uk
Amur tigers - 1 male, 1 female.

Chester Zoo, Upton-by-Chester CH2 1LH.
Signed on A41 Chester Road. Or follow brown Chester Zoo signs from M56, Junction 14, or Junction 12 on M53.
Tel. 08700 664242. Web www.chesterzoo.org
Amur tigers - 1 male, 1 female.

Dublin Zoo, The Zoological Society of Ireland, Dublin 8.
Phoenix Park, Zoo Road, off Wellington Road.
Tel. 00353 16771425. Web www.dublinzoo.ie.
Amur tigers - 1 male, 1 female.

Edinburgh Zoo, Corstorphine Road (A8) Edinburgh EH12 6TF.
4 miles west of city centre.
Tel. 0131 3349171 Web www.edinburghzoo.org.uk
Amur tigers - 1 male, 1 female, 3 cubs.

Linton Zoological Gardens, Hadstock Road, Linton, Cambridgeshire CB1 6NT.
10 miles southeast of Cambridge on B1052, just off A1307. M11 Junction 9 northbound or Junction 10 southbound.
Tel. 0223 891308. Web www.lintonzoo.co.uk
Amur tigers - 1 male, 2 females.

Marwell Zoological Park, Colden Common, Winchester, Hampshire SO21 1JH.
8 miles southeast of Winchester, off B3354 and M3.
Tel. 01962 777407. Web www.marwell.org.uk
Amur tigers - 2 males, 2 females.

South Lakes Wild Animal Park, Crossgates, Dalton-in-Furness, Cumbria LA15 8JR.
On the A590. On M6 follow brown signs from Junction 36 until brown and white elephant signs take over.
Tel. 01229 466086. Web www.wildanimalpark.co.uk
Amur tigers - 1 male, 1 female.

Whipsnade Wild Animal Park, Dunstable, Bedfordshire LU6 2LF.
5 miles south of Dunstable, 10 miles southwest of Luton.
Tel. 01582 872171. Web www.whipsnade.co.uk
Amur tigers - 2 males, 3 females.

Woburn Safari Park, Woburn, Bedfordshire MK17 9QN.
12 miles east of Milton Keynes. Take exit 13 from the M1 or leave A5 at Hockcliffe for the A4012.
Tel. 01525 290407. Web www.woburnsafari.co.uk
Amur tigers - 3 females.

NOTES

Chapter One

[1] Well-written non-specialist books about tigers in general that have been found useful in preparing *Amur Tiger* include the following; there are no doubt many others.

Courtney, Nicholas. 1980. *The Tiger: Symbol of Freedom*, London &c: Quartet Books.

Mountfort, Guy. 1973. *Tigers*, Newton Abbot: David & Charles.

Sankhala, Kailash. 1978. *Tiger! The Story of the Indian Tiger*, London: Collins.

Schaller, George & Selsam, Millicent. 1969. *The Tiger: its Life in the Wild*, Kingswood & London: World's Work.

Thapar, Valmik. 1992. *The Tiger's Destiny*, London: Kyle-Cathie.

[2] Prynn (2003).

[3] Potapov (1976).

Chapter Two

[1] Newly translated from Arsenev (1994), pp.151-157. Dersu is described in the supplement at the end of the chapter.

[2] Arsenev's main task on his expeditions was to explore and survey the Sikhote-Alin Mountains. He was the officer in charge of a small detachment of soldiers, who took care of everything, leaving Arsenev free for cartographic surveying and writing up his observations.

[3] The Li-Fudzin - a Chinese name - is now known as the River Pavlovka; it is on the western macroslope north-west of Olga, and is a right tributary of the Ussuri.

[4] Taiga: normally in Russian the word 'taiga' is used for the northern coniferous forest, but here in the Russian Far East it is applied to all the forest, whether deciduous ('Manchurian') or coniferous ('Okhotsk'). For further details, see chapter 3.

As an English word, 'taiga' is stressed on the first syllable - which is unfortunate in a book about the tiger (pronounced almost identically in England, but not in Scotland). In Russian it is stressed on the second syllable - *taiga*. Perhaps it would be a good idea to 'go Russian' while reading the book, especially if out loud.

[5] Gol'd: the old name for the Nanay people. In the translations some of the old names used by Arsenev for local people and rivers are retained, to help create the atmosphere of the time, before the area had become russianised. In Russian the word 'Gol'd' is pronounced with a palatilised 'l' and unvoiced 'd': in other words the 'l' is like 'l' in 'lute', where the 'u' is pronounced like 'yew'; and the 'd' is pronounced 't'.

[6] Bipod: two poles, joined at the top, for supporting a Berdan rifle when being fired.

[7] Me: the native people of the area mainly spoke a simplified and uninflected pidgin Russian. For 'I' (Russian '*ya*') they would say 'my' (Russian '*maya*'), for 'we' (Russian '*my*') - 'our' (Russian '*nash*'), for 'he' (Russian '*on*') - 'his' (Russian '*ego*', pronounced 'yevo'). It is rather like 'me' instead of 'I' in some forms of English-based pidgin.

[8] Bracken. This is the same species - *Pteridium aquilinum* - as is found very commonly in Britain and the rest of Europe, and indeed all across Russia to Kamchatka - with the exception of desert, steppe and the polar regions. The Russian name is '*orlyak*', which (like '*aquilinum*') refers to the supposed resemblance of the frond to an eagle's wing. Interestingly the plant is regarded as quite useful in Russia, unlike in Britain, where it is a weed that ruins hill pastures and heathlands. For instance, the young shoots can be eaten, after suitable treatment to remove poisons, and are sold in Japanese markets; the rhizome contains c.46% starch, which makes it useful for producing glue or brewing beer; they are used as a substitute for soap in French villages; they have astringent properties and can be used as a weak vermifuge; they are readily eaten by pigs both fresh and as silage, but are poisonous to other stock; they make excellent bedding, and improve the quality of manure; the fronds have a characteristic odour and antiseptic properties, deterring insects; they are used for wrapping fruit and vegetables; the ash contains potassium salts, making it

useful in glass manufacture and soap making. Komarov & Ilin (1934).

If bracken is so useful, one might ask why, in Britain, it is treated as a weed rather than a valuable crop.

[9] *Izyubr* or Manchurian Red Deer *Cervus canadensis*. Known in N.America as 'wapiti' or (presumably originally by mistake) 'elk'. In the Russian Far East the *izyubr* was formerly much hunted for the antlers in velvet.

The earlier name *C.elephas* signified that this animal was regarded as belonging to the same species as the European Red Deer, but it is now recognised as belonging instead to the same species as the Wapiti *C.canadensis* of North America. So, the name 'red deer' is no longer accurate. In this book the Russian name '*izyubr*' will be used, in preference to the North American native name 'wapiti'.

Eastern Roe deer *Capreolus pygargus*: widely distributed through southern Siberia, replacing the European roe deer, which is slightly smaller and has a higher-pitched voice. They are not always recognised as separate species.

[10] Salt licks. Herbivores such as deer often feed almost exclusively on vegetation which is poor in mineral nutrients. To supplement their diet they must find alternative supplies of minerals, hence the importance of mineral-rich soil or rock exposures, which form natural salt licks. Where these are lacking, game managers frequently provide artificial salt licks to help maintain high populations of game species. This and other techniques to promote game populations in the wild are known in Russia as 'biotechnics'.

[11] Northern eagle owl, *Bubo bubo*: a very large owl with well developed ear tufts; found across most of the former Soviet Union as far north as St Petersburg, also most of Europe and into north Africa; occupies a wide variety of habitats, including taiga, deserts and mountains; call a loud distant 'oo-hoo'. Flint *et al.* (1989), p.171.

Eurasian pygmy owl *Glaucidium passerinum*: a very small owl, not common but found throughout the taiga all across the former Soviet Union, also in Scandinavia and central Europe; call said to be a monotonous repetitive 'dyou, oob-oob'. Flint *et al.* (1989), p.177.

One eagle owl weighs the same as c.37 pygmy owls.

[12] The main sources for the section on Arsenev are Mesitov (1994) pp.460-5, and Stephan (1994).

[13] Arsenev's major expeditions were in 1902, 1906, 1907 - all in the southern Sikhote-Alin; 1908-10 - northern Sikhote-Alin; 1918 - Kamchatka; 1922 - Okhotsk Sea coast; 1923 - Komandor Islands, east of Kamchatka; 1927 - exploring a route between Khabarovsk and Sovetskaya Gavan.

[14] Books by Arsenev. The various expeditions were described in the following books: 1902 & 1906, in *Po Ussuriyskomu krayu* (1921); 1907, in *Dersu Uzala* (1922 or '23); 1908-10, in *V gorakh Sikhote-Alinya* (1937); 1927, in *Skvoz' taygu*. The first three expeditions were described in various editions with the title *Dersu Uzala* or *V debryakh Ussuriyskogo kraya*, and the 1908-10 and 1927 ones in a 1949 edition entitled *Skvoz' taygu*. The bibliography is confusing.

Chapter Three

[1] Enisey: both 'e's are pronounced as in 'ye' in 'yes'; the stress is on the last syllable.

[2] There is, incidentally, a line of submerged mountains - the Emperor Sea Mount Chain - stretching south from near the Komandor Islands, and these eventually link up with Hawaii via the Hawaiian Ridge. They are thought to be due to volcanoes being punched through the oceanic crust periodically as it glides over a 'hot spot' in the more fluid layer below. Given a good atlas to see these features, it is difficult to imagine how anyone could fail to become interested in the physical geography and geology of the north Pacific area.

[3] 'Sikhote-Alin' is Manchurian for 'ridge separating two river basins' (Arsenev 1994, p.261, note 66). Arsenev also notes that the local Chinese modified the name to 'Si-kho-ta-Lin', meaning 'pass of the great westward rivers' (*ibid.*, p.149). Marine terraces in the Sikhote-Alin Mountains suggest that the massif was in fact an island in the recent geological past (Suslov 1961, pp.332-3).

[4] Caspian tiger: see chapter 1, 'Caspian Tiger'.

[5] Climate data from Borisov (1965).

[6] People living along the River Amur tune in to daily water-level bulletins with particular interest and anxiety in August. Stephan (1994) p.11.

[7] The phenological difference between east and west macroslopes can be up to 20 or even 30 days (Arsenev 1949, p.305). [8] Arsenev (1994) p.333.

[9] Exploding dittany *Dictamnus dasycarpus*. One of Arsenev's men sat down beside a bush and lit his pipe, whereupon the air around the bush exploded into flame. They found this so amusing that they went around striking matches near other dittany bushes, until Arsenev ordered them to stop, since they were using up the precious matches. The flowers give out such large quantities of ethereal oils that, when there is no wind, the air becomes inflammable. Arsenev (1986) p.118.

[10] Ginseng *Panax ginseng* (formerly *P.schinseng*). Natural range North-east China, North Korea, Primore and Khabarovsk Region; cultivated elsewhere. The North-American ginseng is *P.quinquefolium*, growing wild from Quebec and Manitoba south to the Gulf of Mexico coast. For medicinal use ginseng is prepared as an alcohol infusion of the root, or in the form of powder or tablets; the Chinese make a stimulating tea from it. The Chinese have long regarded it as a panacea (though usually for prophylactic rather than curative use); it is indeed rather non-specific in its effects, having a normalising effect in many pathological conditions and improving mental performance, including learning, memory and sensory awareness. It is thought to act through increasing the brain's ACTH activity. A spiny bush, the so-called Siberian ginseng *Eleutherococcus senticosus*, which grows in the same forests, is said to be a complete substitute for ginseng. *Encyclopedia Britannica, Great Soviet Encyclopaedia*, Court (2000).

[11] Yedo, Yezo or Hondo Spruce *Picea ajanensis* or *jezoensis*.

[12] Arsenev (1994) p.449. Old trees may be up to 300 years old (Suslov 1961, p.348).

[13] The name '*kedr*' is also used for the closely related and similarly five-needled Siberian pine *P.sibirica*.

[14] This section is largely based on Stephan (1994) and *Great Soviet Encyclopaedia*.

[15] Stephan (1994) p.2.

[16] Primore, or Primorski Kray, can be translated as 'Maritime Region'. 'Primore' is stressed on the 'o', and the final 'e' is pronounced like 'ye' in 'yes'.

[17] Khabarov. The town Khabarovsk, originally a military post Khabarovka, was named after him.

[18] The Old Believers were members of the Russian Orthodox Church who refused to accept reforms introduced in the mid 17th century by the Muscovite Patriarch Nikon.

[19] Some Russians performing notable services to the state were honoured with the privilege of attaching a geographical name to their family name. For example, a well

known family of Russian naturalists - the Semënov-Tyan-Shanskis - acquired their triple-barrelled name in honour of Petr Semënov's exploration of the Tyan Shan Mountains in Central Asia. Similarly Muravëv's second barrel commemorates his expansion of Russian power on the River Amur.

[20] Baykov, Nikolay Apollonovich (1872-1958). Born in Kyiv, the capital of Ukraine, the son of a landowner. After schooling in Kyiv and St Petersburg he entered the army, serving first at Tbilisi in Georgia with an infantry regiment commanded by Grand Prince Nikolay Mikhaylovich, who introduced him to the well known naturalist G.I.Radde. Baykov was trained by Radde in taxidermy, and travelled widely in the Caucasus Mountains. In 1901 he was posted to the Chinese Eastern Railway, at a time when it still passed through a large area of virgin forest. He hunted, conducted surveys, and sent zoological collections back to the Academy of Sciences in St Petersburg.

In 1920, during the Civil War (1918-22), he and his family moved to Egypt and Baykov spent two years in Africa, India and Indo-China. In 1922 he returned to Harbin, and migrated to Australia two years before his death.

[21] Sikhote-Alin Meteorite. The largest iron meteorite (mass c.70 tons) to be observed during its fall. It landed in the western spurs of the Sikhote-Alin on 12th February 1947. The bolide was visible from Khabarovsk and Primorski regions over a radius of about 400 km, and left a dusty trail which remained visible for several hours; after the bolide disappeared, booms, rumblings and a roaring sound were heard, and the ground and buildings were felt moving in some areas. Shattering in the atmosphere into thousands of fragments it fell as a shower over 3 square kilometres; 24 meteoritic craters with diameter 9-26 m were found, also 98 penetration funnels (diameter 0.5 - 9 m), and 78 alveoles (diameter less than 0.5 m). The larger fragments, up to several tons each, shattered on impact with cliffs; 27 tons of fragments had been collected by the mid 1970s. *Great Soviet Encyclopaedia.*

[22] Zapovedniks. The history of the whole zapovednik system is described in Shtilmark (2003).

[23] South of or level with Khabarovsk are eight zapovedniks, listed below in order of foundation. Approximate areas are in thousands of hectares (K ha), and the location of each is indicated. Data from Shtilmark (2003) Appendix 1, and D̦zhkin (1989) pp.242-6.

- c.1925, Kedrovaya Pad (meaning Korean-pine Valley), 18 K ha, in hills c.25 km west of Vladivostok across the bay.
- 1932, Ussuriyski (formerly Suputinski), originally 17 K ha, now 40 K ha, in western foothills of Sikhote-Alin, 60 km north-east of Vladivostok.
- 1935, Sikhote-Alinski, originally 1.7 million ha, now 408 K ha, on coast and inland from Terney. A brief history of the zapovednik and its tigers is given at the end of chapter 14.
- 1940, Lazovski (formerly Sudzukhinski; now formally named after Kaplanov), originally 339 K ha, now 120 K ha, dissolved 1952 and re-established 1957, on coast and inland c.150 km east of Vladivostok.
- 1963, Bolshekhekhtsirski, 45 K ha, in hills on southern outskirts of Khabarovsk.
- 1978, Far-East Marine, 64 K ha, islands and sea near Posyet, west of Vladivostok.
- 1990, Khankayski, 38 K ha, part of L.Khanka and its surroundings; important for water birds.
- 1994, Botchinski, 267 K ha, on coast due east of Khabarovsk.

[24] Shtilmark (2003) pp.99-120.

[25] Goral *Nemorhaedus caudatus*. A short-necked goat-like animal with long and luxuriant reddish or grey fur, especially in winter. There is a black stripe along the back, and the tail is black and white, and quite long. The short horns point backwards. Mostly found on slopes near the sea. See also chapter 7, note 20.

Chapter Four

[1] This is a translation of chapter 'Napadeniye tigra' in Arsenev *Dersu Uzala* (1994), pp.441-8. One paragraph (at beginning) omitted.

[2] Yurt: a round or sometimes square hut, usually with lattice walls and conical roof, the frame covered with skins, bark or canvas; a common type of native housing in Central Asia, the Russian Far East, and the Far North.

[3] Fanza: a substantial hut built to a Chinese design, rectangular in plan, with vertical plank walls, windows, and pitched roof, rather like a Russian *izba*.

[4] Sidungou: the name means 'rocky eastern valley'.

[5] *Taza*: a half breed with one parent Chinese and the other from one of the native peoples of the region.

[6] Sable *Martes zibellina*. A medium-sized (up to 1.8 kg) member of the family Mustelidae, which includes the weasels, stoats, martens and badgers. It has luxurious fur, which can be almost any colour from straw to dark brown. Patchily distributed through the taiga zone between the Urals and Pacific coast; formerly much more widely distributed - including northern Scandinavia until the mid 17th century - but exterminated in many areas by intensive trapping.

[7] Rings. The most common kind of ring - or halo - is due to thin high clouds (eg cirrostratus) consisting of ice crystals, which refract light from the sun or moon to create a ring of light at 22 degrees from the sun or moon. A corona is a similar phenomenon, but due to clouds consisting of water droplets.

[8] This storm is described in Arsenev's book about his 1902 expedition: Dersu saves their lives by building a kind of haystack out of reeds, inside which they crawl to escape the hurricane (Arsenev 1986, pp.43-49). This episode was portrayed memorably in Kurosawa's film.

[9] Wild Pig *Sus scrofa*, also known as Wild Boar. Found throughout most of mainland Europe (including England, recently naturalised: see Goulding, 2003), European Russia north to the White Sea, southern Siberia and the Russian Far East; important prey for tigers.

[10] Wolverine *Gulo gulo*. Belonging to the same family (Mustelidae; see note 6 above) as the sable, it somewhat resembles a badger *Meles meles* but is rather larger (up to 30 kg); found all through the northern Eurasian taiga zone from Scandinavia to Kamchatka and beyond. It has a reputation for strength, endurance and ferocity out of all proportion to its size, and is much disliked by trappers for spoiling their catch.

Chapter Five

[1] Taksami (1982). [1] Udege: also known as Udeghs, Udegey, Udegei or Udegeiski.

[3] Aramilev (1961). Other examples of native beliefs about and attitudes to tigers will be found in the even-numbered chapters of this book.

[4] Forsyth (1992). [5] Levin & Potapov (1964).

[6] *izyubr*: see chapter 2, note 9.

[7] Arsenev (1986) p.184. Sankhala (1978, p.64) also comments on Indian tigers imitating the 'ponking' call of sambar deer.

[8] Arsenev (1949) pp.188-9. [9] Levin & Potapov (1964).

[10] *honghutzi* or *hong huzi*: the name given to these bandits means literally 'red-beards', and is possibly derived from a Manchu or Chinese name for the 'red-bearded' Russians who terrorised the Amur natives in the 17th century (Stephan 1994, pp.64,73).

[11] Kolarz (1954).

[12] Committee of the North: the Committee for Assistance to Peoples of the Northern Borderlands was set up under the Revolutionary Committee of the Far East in 1924 by Gamarnik (Stephan 1994, p.178).

[13] Inspection Tiger. Following the establishment of antipoaching patrols (Amba Teams) in 1992-93, the Ministry of the Environment set up Inspection Tiger in 1994 with a view to protecting rare and endangered species, including the tiger. By 1998 there were 8 teams, each typically consisting of 4 members equipped in military style, with radio communications, and with the power to stop and inspect vehicles, confiscate weapons and seize meat or skins of rare species. The scheme has largely been funded by WWF and other foreign donors. Inspection Tiger works with other official security organisations, but has been hampered by the reluctance of the police and courts to punish wildlife crime adequately.

The Tiger Response Team was formed from members of Inspection Tiger assisted by scientists from the Amur Tiger Project (see Supplement to chapter 14) in the winter of 2000-01, as a result of some difficult human-tiger conflicts. In some cases tigers had to be destroyed.

[14] Galster *et al.* (1996). [15] Beltran (2000).

[16] Arsenev (1986) pp.142, 167. It is possible that these shrines and inscriptions were associated with the cult of Guan Yu (also known as Kuan Yü, Kuan Kung, Guan Gong, Guan Di, Guandi, Kuan Ti, Lord Guan and Wu Ti).

The Han dynasty lasted from 206 BC to AD 220, and was followed by the Three Kingdoms period before China was reunited in 280. Guan Yu was a military leader for the kingdom of Shu, and was eventually captured and executed in 219. Subsequently, in the 14th-century *Romance of the Three Kingdoms*, Guan Yu was portrayed as a loyal and courageous general who resisted the evil machinations of Cao Cao (the characters have re-emerged in some computer games and a TV serial). In 1594 a Ming-dynasty emperor raised Guan Yu to the status of Guan Di, the Chinese god of war (Wu Ti) and protector of China and its people, especially the weak and oppressed. Many temples were erected for his worship, and the cult spread to Korea in the 17th century. Guan Di is often portrayed with red face and flowing beard, and in so far as he is associated with any animal it is usually the horse as a symbol of service. The background to the shrines described by Arsenev requires more research, especially the link with tigers. (I am grateful to Dr Hongxing Zhang of the National Museums of Scotland for suggesting the link with Guan Yu.)

[17] Sowerby (1923). [18] Stephan (1994).

[19] British Museum: The Arts of Korea Gallery, 2002.

[20] Sowerby (1933). [21] Arsenev (1994) p.209.

[22] Mention of Amur tigers preying on dogs will be found in several of the stories in the even-numbered chapters of this book. Interestingly, Sankhala (1978, pp.112,114) notes that Indian tigers usually avoid eating dogs. This may be simply a question of general abundance of food, dogs not being preferred and so ignored where other prey is easy to find. If however it is a consistent behavioural difference between Indian and Amur tigers, it may be explained by the tigers' ecological relations with wild relatives of the domestic dog - namely, for the Amur tiger: wolf, fox, Indian wild dog (formerly), and raccoon-dog: and for the Indian tiger: wolf and jackal *Canis aureus* among others. There is thought to be a negative correlation between the abundance of tiger and wolf in the Russian Far East: see the end of chapter 11. [23] Matyushkin (1996).

[24] Smirnov S. (1881), quoted in Matyushkin *et al.* 1978.

[25] Loukashkin (1938).

[26] Baykov (1936). For details about Baykov, see note 20 to chapter 3.

[27] Matyushkin *et al.* (1980). A description of how tigers were captured is given in chapter 12, below. [28] Thiel (1957).

[29] Arsenev (1994) pp.318-19. [30] Bannikov (1974).

[31] Sowerby (1934). [32] Matyushkin (1998).

Chapter Six

[1] Translated from Arsenev (1949) (pp.191-204), which describes his expedition in the Sikhote-Alin Mountains in 1908-10. The story takes place on the River Anyuy, north-east of Khabarovsk.

[2] Apart from sharpening its claws on the tree, the tiger is probably also marking its territory, and by the height of the claw marks above the ground it is showing how large an animal it is. [3] yurt: see chapter 4, note 2.

[4] The crossbow has a long history; it was used by the Chinese as early as the 4th century BC. There was even a repeating crossbow, with a magazine to contain 10 darts, which could be fired in 15 seconds; this was still being used in the 1894-95 war between China and Japan (Payne-Galllwey, pp.237-42). Crossbows were also used by the Canadian inland Inuit (Mowat 1952, pp.234, 237).

[5] *chumiza*: a cereal grain (from the grass *Setaria viridis*).

[6] In cold weather, blood vessels next to the skin contract, reducing the amount of blood at the surface of the body and so reducing heat loss. Rubbing the skin with snow encourages the blood vessels under the skin to dilate slowly, in response to physical irritation. If instead dilation occurs in response to rising air temperature it may be too rapid, with consequent damage similar to chillblains and frostbite. (Information from D.G.Boddington.)

Chapter Seven

[1] This quotation comes from the main source of information for this chapter - Formozov (1948). Other sources are Geptner (1971), Kaplanov (1948), Malyshev (1971), and Matyushkin (1993).

[2] 'Kaplanov' is stressed on the second syllable, and the last letter is not voiced; hence 'kaplanof'.

[3] Similarly '*lev*' is unvoiced on the last letter, and the vowel is palatalised, ie it begins with a 'y' sound as in 'yes'; hence 'lyef'. His patronymic is 'Georgiyevich', meaning 'son of George', because his father's given name - 'Gazaros' - is equivalent to 'George' in Armenian. (Information from Feliks Shtilmark.)

[4] Vadim Vadimovich Rayevski (1904-47): a field biologist whose most important scientific work was a study of sable at Kondo-Sosva Zapovednik. He died of tuberculosis in his early forties, probably exacerbated by the harsh conditions of life at the zapovednik. For further details see Shtilmark (2002, 2003).

[5] '*KYuBZ*' stands for '*Kruzhok yunykh biologov Moskovskogo zooparka*' (Circle of Young Biologists of Moscow Zoo). The club - then only recently founded, and still running today - achieved fame as providing encouragement to many children who later became well known biologists. A similar club is organised at the University Zoology Museum in Moscow, and their activities include expeditions to various parts of Russia.

[6] Alexandr Nikolayevich Formozov (1899-1973): prominent animal ecologist, who dominated the field in Russia in his time. His best known scientific work concerned the effects on animals of snow cover, and among his more popular books was *The Tracker's Companion*, which described animals' spoor and other signs. Like Kaplanov, as a boy Formozov enjoyed extended trips into the forest to learn about nature - in his case using the countryside around Nizhni Novgorod on the River Volga.

[7] Acclimatisation and introductions: for further details see Shtilmark (2003) pp.71-3.

[8] Elk domestication: a chapter in Smirin & Smirin (1999) describes elk observed at an elk farm. [9] Formozov (2000) p.174.

[10] Matyushkin (1993) p.10: letter to Formozov of 18th December 1934, from Usaykov on the River Demyanka.

[11] Load. In his book *Tiger, Izyubr, Elk*, Kaplanov describes in detail how they cut their loads to the bare minimum when making winter treks to survey tigers. Even in the depths

of winter they would dispense with the usual sledge, heavy tent, portable stove, normal fur sleeping bags, binoculars, telephoto lens, washing kit and spare underwear. Everything was carried in backpacks, each weighing up to 25 kg.

A typical load would include a light axe, saw, lightweight tent (flysheet only), saucepan, mug, spoon, camera, rifle, compass, clock or watch, knife, file, chisel, whetstone, notebook and pencil, awl, needle and waxed thread, and pieces of chamois leather to mend footwear. Food was chosen for maximum energy per kilogram - rusks made of 30% wheat flour, honey, butter, rice, bacon, sugar, cocoa and alcohol. Meat was obtained *en route* from remains of tiger prey. Clothing consisted of jacket and trousers, warm underwear, woollen sweater, fur cap, and fur gloves. Footwear was native boots *(ologi)* made from chamois leather (from elk or *izyubr*) lined with a particular species of plant with soft stems. Native Udege skis were used in deep snow - about 2 m long and 13 cm wide, made from ash and lined underneath with skin from elk leg (with the hairs pointing backwards). Dogs were often taken along, partly to serve as a mobile emergency food supply, or to pull a sledge if one had to be made on the way. The dogs were fed on remains of prey killed by tiger, wolf or yellow-throated marten. On one occasion, owing to unforeseen circumstances, food ran out and no meat could be found; so the people had nothing to eat for three days. It is on such occasions that the dogs may have to be sacrificed.

[12] A rifle trap is described in chapter 6. Such traps are extremely dangerous, since animals or humans other than the intended target can be wounded or killed.

[13] Formozov (2000) p.183.

[14] Pacing gaits. In long-legged animals like the camel, giraffe and gerenuk, the two legs of one side may collide if the left fore and right hind legs move forward simultaneously, followed by the right fore and left hind ones. Some long-legged dogs also use the pacing gait. It is not clear why relatively short-legged and long-bodied animals like the tiger and cheetah use it. (R.McNeill Alexander, letter to *New Scientist* 16 August 2003, p.85.)

[15] Moguchi. Any resemblance to 'Moggy' is purely coincidental. Underlining shows where the word is stressed. [16] Kaplanov (1948) pp.32-33.

[17] It has also been noted in India how deer may remain close to a feeding tiger, apparently quite unconcerned; they presumably 'know' that they are safe, since the tiger will not hunt in the near future (Sankhala 1976, p.54).

[18] Matyushkin (1993) p.14. From various remarks made by Kaplanov it appears that his attitude to killing animals in the zapovednik, even for research purposes, and to the continuation of cub capture for zoos (Kaplanov, 1948) are far removed from the principle of *zapovednost*, on which the zapovedniks were supposed to be run. *Zapovednost* is based on non-intervention management with a view to seeing how natural ecosystems develop without any human interference, the only permitted human activities being scientific monitoring and guarding the reserve. However, especially at this time in Soviet history, there were strong pressures on zapovedniks not only to generate their own funding - and they are facing similar pressures now in the early 21st century - but also to contribute to the national economy. So it is perhaps understandable that Kaplanov, having been strongly influenced by game biologists and living at the time he did, was not too bothered about *zapovednost*.

For further details about *zapovednost* and how far the ideal was realised through the history of the zapovednik system, see Shtilmark (2003).

[19] Formozov (2000) p.183: letter to Formozov of 25th January 1940. If the language seems a bit formal and stilted, it is probably because Kaplanov expected, or at least hoped, that Formozov would read it out to a meeting of fellow zoologists or would publish extracts.

[20] Gorals in the Sikhote-Alin Zapovednik are the subject of chapter 11 (pp.181-193) in Smirin & Smirin (1999).

[21] For an up to date account of protection in the zapovednik, see Laptev (2003).

[22] This kind of tragedy is by no means a thing of the past. While this chapter was being written, news was received of the death in rather similar circumstances of the 39 year old botanist Aleksey Turuta in Pribaykalski National Park on the west side of Lake Baykal. On 26 July 2002 he set off to visit the remotest of the monitoring and experimental plots in the park. He failed to return as expected on 5 August. A search began on 9 August, including Aleksey's wife. The body was found on 5 September; he had been murdered. Even today, the life of an ecologist or conservationist in the remoter parts of Russia - and probably any other part of the world - is not without its risks. Ryabtsev (2003).

[23] Matyushkin (1993) p.12: letter to V.V.Vasilev of 7th June 1936. 'Manchurian bear' presumably refers to the brown bear *Ursus arctos*; the Lord of the Mountains and Forest is presumably the tiger.

[24] Kaplanov's murder. The alternative theory and evidence for it are discussed in Shtilmark (2003) pp.95-96.

Chapter Eight

[1] Translated from Arsenev (1949) pp.176-182.

[2] River Sadomabirani: a small river c.75 km east of Khabarovsk, to the west of the River Khor; it flows into the River Mukhen.

[3] verst, *versta*: an old Russian measure of distance, equal to 1.06 km.

[4] *kasha*: if made of oatmeal it is porridge, but Russians make similar dishes from various other grains and seeds.

[5] Musk deer *Moschus moschiferus*: not a true deer (family Cervidae), but a member of the Moschidae; like a small deer, c.50-60 cm high at the withers and weighing 15-17 kg, with hind quarters rather larger than the fore quarters and no antlers, but instead a small tusk on each side of the upper jaw. The highly prized musk is obtained from a gland in the male.

Chapter Nine

[1] Hemmer (1987), Zdansky (1924).

[2] Seidensticker (1996).

[3] Sunquist *et al.* (1999).

[4] Kitchener (1991, 1999).

[5] Ortolani & Caro (1996).

[6] Geist (1987). Note that there are two kinds of size variation. A tiger population feeding habitually on large prey will be expected to evolve large body size through natural selection, because only large tigers can kill large prey; this is genetic variation. The large body size associated simply with being well fed during development is not genetic.

[7] Hemmer (1987).

[8] For a summary of the 'races' of tiger, please refer to Appendix 2. If different 'races' of tiger are hardly different genetically - and tests have shown no differences between South China and Indochinese tigers (information from Christie, June 2003) - it may justify 'hybridisation' between 'races' to improve the genetic condition of a small population.

[9] If the Sumatran tiger belongs to the same subspecies as the Java tiger, they will be known as *Panthera tigris sondaica*.

[10] The Amur tiger was formerly recognised as *Panthera tigris altaica* Temminck 1844, based on a skin brought from Japan by Siebold and said to have come from Korea (*Fauna Japonica*, 1844): Temminck stated that the specific location was the Pisihan Mountains at 42 degrees N. The type specimen is now at the Rijksmuseum van Natuurlijke in Leiden, Netherlands. Other names that have been used for the Amur tiger are: 1868 *Felis longipilis* Fitzinger (Amur region); 1871 *F.tigris* var.*amurensis* Dode (Ussuri estuary, Khabarovsk); 1904 *F.tigris coreensis* Brass (N.Korea); 1915 *Tigris Mikadoi* Satutin (Korea); 1925 *F.tigris mandshurica* Baikov (N.E. China). Heptner & Sludskii (1992).

[11] Dinerstein *et al* (1997).

[12] Kitchener & Dugmore (2000).

[13] Pocock (1929); Mazák (1967).

[14] Stroganov (1962). It has been pointed out that, despite the snowy winters it has to hunt in, the Amur tiger does not go white in winter: 'the tawny colour and black stripes must be a disadvantage for hunting' (Sankhala 1978, p.23). On the other hand Suslov (1961, p.353) notes that snow cover forms late and disappears early, and may be blown away or covered by leaf litter. Only a few animals in the Russian Far East turn white in winter, mountain hare and stoat being two examples. The sika deer even becomes darker - by losing its spots.

[15] *Phoenix Fund Report,* Vladivostok 2001.

Inbreeding. How has a small population of a few hundred Amur tigers remained healthy and not become inbred? Field studies of Amur tigers undertaken in 1984 and 1985 (Bragin 1986) suggested that, because of their particular type of social interactions, the Amur tiger can exist without the threat of inbreeding depression at lower numbers than, say, the Indian tiger. However, later genetic studies of a small number of tissue samples from different parts of the Russian Far East indicate that Amur tigers are indeed very closely related, and this may present problems of inbreeding in the future (Phoenix Fund Projects 2001). As only 25-30 tigers survived in the Russian Far East in the 1940s the present population is clearly descended from very few individuals.

[16] Korean tiger: known as *Panthera tigris coreensis*, Brass 1905.

[17] Heptner & Sludskii (1992).

[18] Mazak (1967). Baykov at first accepted that there were two tiger subspecies in the region, and the Korean tiger was even given the name *mikadoi*, in honour of the Japanese emperor; later, after years of studying - and hunting - tigers in Manchuria, Baykov concluded that the two forms of tiger were not distinguishable (Loukashkin 1938). A female tiger from North Korea exhibited in the 1960s at Prague Zoo was described as not significantly different from the Russian Far East tigers in the collection (Veselovsky 1967).

[19] Mazák (1981).

[20] Sysoev (1960).

[21] Baykov (1925, 1927).

[22] Information from Smirnov, Sept.2001.

[23] Information from Yudin, Sept.2001.

[24] Vakhreev *et al* (1983).

[25] The tiger was declared India's national animal in 1972 (Sankhala 1978, p.130). Could the same be done at a regional level in the Russian Far East?

[26] Yakutsk: tigers feature in the folklore of the Yakut people, but this is probably due to the fact that the latter immigrated to Yakutia (now known as Republic of Sakha) from further south, in the region of Lake Baykal, bringing their folklore with them (Thubron 2000, p.250). Fossil material of tiger is said to have been found in the Lena basin and some islands in the Arctic Ocean (Sankhala 1978, p.23), but identifying such material is difficult. See also Kitchener & Dugmore (2000).

[27] Mazak (1967).

[28] Sowerby (1923).

[29] Greater Khingan: also known as Da Hingganling or Daxinganshan.

[30] Wang Chi-Wu (1961).

[31] Ma Yiqing *et al.* (1997).

[32] Lesser Khingan: also known as Xiao Hinggan Shan.

[33] Cheng Jizhen (1999).

[34] Matyushkin *et al.* (1980). Manchuria now comprises the Chinese provinces of Liaoning, Jilin (formerly known as Kirin) and Heilongjiang (Heilungkiang). 'Heilong' is the Chinese name for the Amur.

[35] Lu Houji (1987).

[36] *Cat News* 1998, no. 29.

[37] Williamson (1870).

[38] Sludskii (1972).

[39] Wu Xianzhong (1994).

[40] Xiaochen Yu *et al.* (2000).

[41] The survey was sponsored by the UN Development Programme, and was conducted in Yankian Prefecture between February and March 1998. See also *ITAR TASS News Agency* Aug 5 2002.

[42] Matyushkin *et al.* (1996).

[43] Miquelle (1998b).

[44] A zakaznik is a type of nature reserve offering less complete protection than in a zapovednik, either for only a limited period and/or protecting only some species.

[45] Miquelle (1998b).

[46] Matyushkin *et al* (1980).

[47] Attenborough (2002) p.149. David Attenborough has graciously acknowledged that it was a mistake. (But see note 26.)

[48] Kuchler (1948).

[49] Thubron (2000) p.249.

[50] Vasilev & Matyushkin (1982) p.252.

[51] R. K. Maak (1861).

[52] *Interfax News Agency,* May 28, 2001.

Chapter Ten

[1] Translated from Baykov, N.A. *V gorakh i lesakh Man'chzhurii* (In the Mountains & Forests of Manchuria), Petrograd, 1915; reprinted in *Okhotnich'i prostory*, book 3, Moscow: Izd-vo okhotnichey lit-y, 1998. Three paragraphs at the end have been omitted, two of them discussing the use of dogs in hunting tigers.

For biographical notes on Baykov, see chapter 3, note 20.

The story is set in Manchuria near the Chinese Eastern Railway where it passes through mountains between Harbin and the Russian frontier, during the tigers' mating season.

Another of Nikolay Baykov's books - *Big Game Hunting in Manchuria* (English translation, London: Hutchinson, 1936) - contains a chapter called 'The Tigers' Night'. The story is set in December. An old Chinese trapper, Li-Sun - speaking in a hushed voice, an expression of awe-struck reverence on his worn face - explains that 'tigers' night' refers to the tigers' 'winter mating ritual. 'This night, and for the next two nights, is the time of the Tigers' annual festival. They answer the call of their Chief, the Great Van, and come from all over the forest to join him here. They are after blood. No other animal will remain in their vicinity at the time of the Tigers' festival: they go away, leaving the tigers to the undisputed possession of the forest. If the tigers cannot find any other animal to kill, they tear to pieces one of their own community and drink his blood. Only the Great Van does not take part in this ceremony; he looks on from the top of a rock.'

[2] verst, in Russian *versta*: an old Russian measure, equal to 1.06 kilometres.

[3] *rozhdestvo*: literally 'Christmas' or 'Nativity'.

[4] An old Russian measure, one *vershok* = c.4.4 cm; 4.5 *vershok* = c.20 cm.

[5] Indian wild dog *Cuon alpinus*, known in Russian as *krasny volk* (red wolf); resembles a large fox with a short pointed muzzle; no longer resident in Russia, but may wander across the border from China or Mongolia.

[6] *pood*: an old Russian measure, equal to 16.38 kg. This makes the pig c.80 kg. Especially large wild pig are found in the Russian Far East, up to 300 kg.

[7] snack: probably referring to Russians' partiality for suet or other fat, sometimes eaten raw by hunters from a fresh kill.

[8] platform: hunters' and other log cabins in the Russian taiga are generally fitted with a platform (*nary*, or *kany* in the Russian Far East) under which air from the fire can circulate to warm the people sleeping on it.

[9] *arshin*: another old Russian measure, equal to 16 *vershok* or c. 71 cm; so the tiger would have been c.1 m high.

[10] *zubrovka* (or *zubrowka*) is sweet-grass vodka, flavoured with sweet grass or holy grass *Hierochloë odorata*, which is said to be favoured by European bison (*zubr*) (*New Shorter Oxford English Dictionary*, 1993).

[11] 1 *sazhen* = 3 *arshin* or 2.13 m; so the tiger was c.20 m above the author. '*Sazhen*' can be stressed on either syllable. 500 *sazhen* = 1 verst (see note 2).

Chapter Eleven

[1] Ognev (1935).

[2] Tiger calls. 'Tigers produce a wide variety of sounds, from deep roars and growls to the raspberry-like "chuffing" they use to greet each other.' But are tigers hearing the same sounds that we do? Researchers in Nebraska have recently investigated the frequencies of the sounds, and discovered that in many tiger calls a large amount of acoustic energy goes into the lower-frequency sounds - which carry to greater distances and are less likely to be affected by atmospheric conditions and vegetation. A carrying distance of 8 km has been quoted. Tests on an Amur tiger showed some sound being produced in the infrasound range, below 20 hertz; tigers can almost certainly hear these sounds, but we cannot. It is possible that sensitive microphones placed in the forest to detect these frequencies could improve tiger monitoring. (*New Scientist*, no.2393, 3rd May 2003, p.21.)

[3] Gibson (1997).

[4] It is generally accepted that the Amur tiger is territorial, but territoriality has been denied for Indian tigers, where marking (cg with urine) is thought simply to enable tigers to keep well separated (Sankhala 1978, pp.39-41, 206).

[5] Matyushkin *et al.* (1980).

[6] Energy efficiency. This means that, for a given amount of energy expended on making a kill, the food 'reward' is greater if a larger animal is killed. If a tiger spent its whole time feeding on mice, all the energy gained from the mouse-size mouthfuls would probably be spent on catching the next mouse, with none left over to keep warm, move around, or reproduce. An exclusively mouse-eating tiger would soon die.

[7] In 2000 there were c.2500 *izyubr* and c.1200 wild pigs in Sikhote Alin Zapovednik - representing the main prey for some 40 tigers. By contrast in India's Kanha National Park - about half the area of the Russian tiger reserve - there were more than 25,000 chital or axis deer as well as thousands of sambar, muntjac and barasingha deer, nilgai antelope, wild pigs, gaur or wild cattle and other species to support c.100 tigers.

[8] Matyushkin *et al.* (1980).

[9] But see chapter 14, Supplement.

[10] Estimates of tiger diet are based on located kills (Miquelle *et al.* 1996). Assuming that an adult tiger kills at least one deer or wild pig each week it will clearly need more than 50 kills a year. As the extraction rate is about 10% of the ungulate population (hoofed animals), one tiger depends on about 500 large ungulates as a prey base. The 1997 Russian *Strategy for the Conservation of the Amur Tiger* considered 425-510 adult tigers a suitable number for Primorski Region. For this to be maintained, it has been calculated that a stable number of ungulates - wild pig and *izyubr*, roe and sika deer, as well as musk deer, elk and gorals - totalling 190,000-230,000 is required. Approximately 10-16% of these are likely to be taken by tigers annually. Stamotyuk (1998).

See chapter 13, note 3. It has to be admitted that there is a serious credibility gap in maintaining that about 400 tigers are managing to survive on such an apparently inadequate prey base. All of the following are possible: (a) ungulate populations are seriously underestimated; (b) tiger populations are overestimated; (c) tigers have other significant food sources (such as badgers); (d) tiger numbers in the Russian Far East are boosted by immigration from China. (With acknowledgement to Andrew Kitchener.)

[11] Like the tiger, pigs find snow depth of more than 30 cm difficult to traverse (Miquelle *et al* 1999b).

[12] This is the first time such an event has been recorded (Goodrich 2003, or information from J. Goodrich 2003).

[13] Heptner & Sludskii (1972).

[14] The Ussurian Wild Pig *Sus scrofa ussuricus*, Heude, 1888. Boars weigh as much as 160-260 kg, reach 200 cm in length and stand up to 118cm at the shoulder.

[15] Bromley (1964).

[16] Pikunov *et al* (1978).

[17] Poor acorn crops and human hunting pressure led to a drastic decline in the number of wild pig from 30,000-40,000 in the 1960s to c.16,000 in the late '70s and only c.7000 in Primore in the late '80s; they have disappeared from many parts of their former range (Pikunov 1992). [18] Pikunov *et al* (1989).

[19] *Izyubr* (Russian name), Manchurian Red Deer or Wapiti *Cervus canadensis xanthopygus* Milne-Edwards 1867. The North American name is 'wapiti' or, confusingly, 'elk'. Larger than the European Red Deer *C.elephas*, the *izyubr* weighs up to 250 kg and reaches 140 cm at the shoulder.

[20] In the 1970s it was estimated that there were in average years c.22,000 *izyubr* in Primore and about the same number in Khabarovsk Region.

[21] Miquelle *et al* (1999b).

[22] Sika Deer. The largest and most heavily spotted race of the species is found in the Far East - Dybowski's Deer, or Manchurian or Ussuri Sika *Cervus nippon mantchuricus* Swinhoe, 1864. Stags weigh 130-140 kg and reach 109 cm at the shoulder.

[23] The wild sika deer population in Primore increased to more than 2000 by the 1970s and 3000-4000 by the late '90s. Wild sika populations fluctuate widely. Some sika have been released from bankrupt deer farms - previously state farms but some failed during privatisation in the 1990s. Tigers and leopards frequently attack farmed sika in Primore though conservation bodies pay compensation to farmers for losses.

[24] Manchurian, Chinese or Ussuri Roe Deer *Capreolus pygargus bedfordi* Thomas, 1908 are larger and have longer legs and slightly more complex antlers than the European form. They weigh 32-40 kg and attain 80-83 cm in height. In the 1980s there were c.10,000 roe in Primore, with probably a similar number in Khabarovsk Region.

[25] Far Eastern Musk Deer *Moschus moschiferus parvipes* Hollister 1911. They are some 7 to 17 kg in weight, and reach 40-60 cm at the shoulder

[26] Matyushkin *et al* (1980).

[27] Amur or Korean Long-tailed Goral or goat-antelope *Naemorhaedus caudatus raddeanus* Heude, 1894. It represents the most northerly subspecies of the Grey Goral *Naemorhaedus caudatus*. Chapter 11 of Smirin & Smirin (1999) - also published by Russian Nature Press - describes gorals in Sikhote-Alin and Lazovski zapovedniks (with many drawings).

[28] Myslenkov & Voloshina (1989) estimated that some 600-750 gorals were living in Primore in two main populations. The total Primore population remained reasonably stable during the 1990s, at c.600-800 at the end of the decade. A much smaller number is found in Northeast China and in North and South Korea.

[29] Elk *Alces alces*, in North America known as Moose. The Manchurian Elk *Alces alces cameloides* Milne-Edwards, 1867 is the smallest elk subspecies. The antlers have rather little palmation. Bulls weigh 250-400 kg and stand 160-180 cm at the shoulder. The Alaskan Moose *Alces alces gigas* is much larger, weighing as much as 800 kg, up to 2.3 m high at the shoulder, and with huge palmate antlers.

[30] The Ussurian Black Bear *Ursus* (or *Selenarctos*) *thibetanus ussuricus* Heude, 1901 is the northernmost race of the Himalayan or Asiatic Black Bear. It weighs 100-200 kg.

[31] Khramtsov & Zhivotchenko (1983). The same writers found that 45-50 black bears and 10-15 brown bears annually inhabited the zapovednik in 1980-90. There may have been 25,000-35,000 black bears in the Russian Far East in the early 19th century (Chestin & Yudin 1999). As a result of hunting, by the late 1990s the population had declined to 4000-6000 in Primore and southern Khabarovsk Region.

[32] The Far Eastern or Amur Leopard *Panthera pardus orientalis* Schlegel, 1857 is the northernmost subspecies of leopard. Adult males reach 107-136 cm in body length, with a tail 82-90 cm long, and weigh 32-48 kg, exceptionally 60 kg.

[33] Heptner & Sludskii (1972).

[34] Abramov & Pikunov (1976). [35] Information from Korkishko, Sept. 2001.

[36] By the 1980s leopards had disappeared from two of these areas, surviving in only one small c.260 x 40 km strip, mostly in Khasanski District. along the Chinese border in the extreme south-west of Primore.

[37] Xingjia Yang & Jinsong Jiang (1996). [38] Pikunov *et al* (1999).

[39] Manchurian Brown Bear *Ursus arctos lasiotus*. In the 1990s there were probably c.2500-3000 in Primore and over 5000 in southern Khabarovsk Region (Chestin & Yudin 1999). In China's Heilongjiang Province there were an estimated 500-1500 wild brown bears in the 1990s, and fewer in Jilin Province. [40] Gromov & Matyushkin (1974).

Chapter Twelve

[1] Translated from Zhidenkov (1966). The story is set in the hills to the east of Lake Khanka.

[2] *shatun*: Russians have several names for brown bears. As well as the word 'bear' itself - *medved*, derived from *mëd*, meaning 'honey' - the animal is often referred to as *shatun*, meaning 'loafer, idler', or *kosolapy*, meaning 'clumsy, pigeon-toed'. A common pet name, especially for a teddy bear, is *Mishka*, a diminutive of *Mikhayil* or 'Michael'.

A *shatun* is specifically a bear which does not fatten up in the autumn and then undergo the usual hibernation in a den, but instead wanders through the forest during the winter, eating anything it can find; these are dangerous animals.

[3] Volchok: the dog's name, *volchok*, means 'spinning top'; but there may be connotations of *volk* ('wolf') and *volchonok* ('wolf cub').

[4] suet, in Russian *salo*: fat from under the skin or from the body cavity of pig, elk, or other animal; a delicacy among hunters, eaten either raw or fried (from pig, fried, is tastiest - according to Felix Shtilmark).

[5] *rogulina*: this forked stick, about 3 m long, is best cut from oak, birch, or other broadleaf tree that branches at a fairly acute angle. It is used for pinning down a tiger by its neck or leg, and must therefore be strong. The Russian word *rog* means 'horn' or 'antler'.

[6] 'Siberian stove' or *nodya*: while small metal stoves were frequently carried by even quite modest camping expeditions in Russia, since a good source of heat is essential for survival in the taiga during winter, 'Siberian stove' is a whimsical description of a camp fire consisting of two large logs, laid along the ground side by side: the fire burns between and under them, thus being largely enclosed and generating a good heat.

[7] *portyanka*: when socks are not available, or not issued (eg in the Russian forces), or sometimes for preference, Russians wrap a rectangle of cloth around their feet, to serve as a sock. [8] Solontsovy: a place name probably referring to a salt lick (*solonets*).

Chapter Thirteen

[1] Kaplanov's life and work are described in chapter 7.

[2] Caspian tiger: see chapter 1.

[3] See table ('Estimated numbers ...') on next page.

[4] Sysoev (1960). [5] Spitsin *et al.* (1987). See also p.119.

[6] Matyushkin *et al.* (1980). [7] Dunishenko (1985).

[8] Eg Pikunov (1988b). If tigers' capacity to cover great distances is underestimated, they might be counted more than once.

[9] *New Scientist* 30 November 1996, p.5 and 12 July 1997, p.18.

[10] Heptner & Sludskii (1972).

[11] In a vast country of diverse habitats, there were in the 20th century no less than 5 of the 8 usually recognised subspecies or regional populations of tigers within China's national frontiers, often existing in large numbers over much of the country. These were the Amur, Caspian, Indian, Indo-Chinese and South China tigers (for more details see Appendix 2). Today there is hardly a viable tiger population surviving anywhere in China.

Estimated numbers of Amur tigers in the Russian Far East 1947-96

Year	Primore	Khabarovsk Region	Total	Source
1867	-	-	c.1000[a]	Matyushkin (1996)
1947	14-16	11-14	25-30	Kaplanov (1948)
1958-59	55-60	23	78-83	Abramov (1970)
1969-70	110-134	20	130-154	Yudakov & Nikolayev (1987)
1978-79	172-195	34	206-229	Pikunov (1988b)
1984-85	285-293	42-50	327-343	Pikunov (1990)
1989	360-370	60-64	420-434	Bragin & Gapanov (1989)
1995-6	-	-	415-475[b]	Matyushkin *et al.* (1996a)
c.1996	-	-	250[c]	
1996 +	-	-	400-450	Miquelle (2000)[d]

[a] estimate for the whole of Russian Far East.

[b] based on winter tracking; *New Scientist* 30 Nov.1996, p.5.

[c] extrapolated from estimate of 12 tigers in Lazovski Zapovednik, using sniffer dogs; *New Scientist* 12 July 1997, p.18; *The Times* 10 July 1997. See Supplement to chapter 14.

[d] Since 1996 the Amur tiger population in Russia has been monitored from 16 survey stations and their numbers, though decreasing in some areas, have increased in others and are reported as stable at 400-450.

Note: serious doubts are now being cast on the methods used for estimating tiger populations in the past, so that the figures quoted here may be inaccurate and misleading.

[12] Miquelle *et al.* (1995b). [13] Goodrich *et al.* (2002).

[14] *Cat News* no.29, Autumn 1998. The foreign organisations mentioned included IUCN Cat Specialist Group, Zoological Society of London (as part of the 21st Century Tiger Alliance) and WWF. [15] Oshmarin & Pikunov (1990).

[16] BBC Radio 4 Report, 8th February 1987.

[17] Massey Stewart (1992). [18] Zhivotchenko (1985).

[19] Dunishenko (1985). [20] Bragin & Gapanov (1989).

[21] Prices are given in US dollars.

[22] Pikunov (1994). See also TRAFFIC Bulletin, vol.15, no.1, 1994.

Chapter Fourteen

[1] Translation of Smirnov (2000), pp.32-42, 'Tiger Monitoring in the Zapovednik'.

[2] Slightly different figures are given at the end of the chapter.

[3] Monitoring in winter: the Russian phrase is 'along the white path' (*po beloy trope*), meaning following spoor and other signs in the snow; when the snow has gone, monitoring is 'along the black path' (*po chërnoy trope*).

[4] Siberian Weasel or Kolinski *Mustela sibirica*. The name 'Siberian weasel' is not satisfactory since *M.sibirica*, at 650-820 g, is larger than a European stoat *M.erminea* (up to 260 g), let alone the small European weasel. The Kolinski's fur is of an orange tint. Found throughout Siberia from the southern Urals to the Sea of Okhotsk and the Russian Far East.

[5] Mountain Hare *Lepus timidus*: found all across northern Eurasia, including Ireland and Scotland. Generally turns white in winter except for the tips of its ears.

Manchurian Hare *L.mandshuricus*: in Russia, confined to the Russian Far East (Ussuri and Amur valleys, and coast of Sikhote-Alin Mountains); a small hare; fur brown above, white below. There is considerable confusion in Russian works about the scientific name, probably reflecting uncertainty about the taxonomy.

[6] American Mink *Mustela vison*: introduced to the Russian Far East as a valuable fur animal. The native European Mink *M.lutreola* extends only a short way into western Siberia.

[7] Otter *Lutra lutra*: found all across northern Eurasia from western Europe to the Pacific, and as far north as the Arctic Ocean in Fennoscandia and Chukotka.

[8] Yellow-throated or Indian Marten *Martes flavigula*: found in Russia only in the Russian Far East; a striking animal, with yellowish body but dark brown tail, legs and head except for the white chin; when leaping through the tree tops it is said to resemble a monkey (Dinets & Rotshild 1996, p.106).

[9] The Russian for 'wood goblin' or 'wood devil' is *leshi*. One of Arsenev's dogs was called Leshi.

[10] Chronicle of Nature, *letopis' prirody*: a set of phenological and other observations maintained year after year in each zapovednik, and largely standardised across all 100 zapovedniks in Russia (and formerly in many more throughout the Soviet Union). This forms a valuable long-term series of data essential for understanding how the biosphere works, particularly in conditions of climate change. See Shtilmark 2003, Harper *et al.*.

[11] Kucherenko 1970.

[12] Tiger censuses were organised in the winters of 1956-57 and 1958-59 (K.G.Abramov and V.K.Abramov), 1969-70 (A.G.Yudakov and I.G.Nikolaev), 1977-78 and 1978-79 (D.G.Pikunov and V.K.Abramov), and 1983-84 and 1984-85 (D.G.Pikunov and A.I.Bragin). Use of pug marks to estimate sex and age of tigers is not very reliable; see Note 22.

[13] Salkina 1994.

[14] Yudakov & Nikolaev 1987.

[15] Smirnov 1993.

[16] Matyushkin 1992.

[17] Khramtsov 1995.

[18] The Hornocker Wildlife Research Institute has now merged with the Wildlife Conservation Society based at Bronx Zoo, New York.

[19] Information from John Goodrich, July 2003.

[20] Badger *Meles meles*: a large member of the mustelid family, with black-and-reddish striped head (not so striking as the black-and-white head of the European form of the species). It is found all across northern Eurasia from western Europe to the Russian Far East, extending as far north as the White Sea in the west, but north only to the Amur estuary on the Pacific coast.

[21] Information from L.Kerley, July 2003. Using scent is a much less intrusive method than snaring tigers for radio-collaring. The latter method subjects the animals to shock, however benign the leg holds may be.

[22] Information from L.Kerley, July 2003. Kaplanov himself noted an example of poor correlation between pug mark and body size (see chapter 7). It remains the case, though, that pug marks with plantar pad over 10 cm wide are likely to indicate adult males.

Chapter Fifteen

[1] Christie (1995a, 1998, 2000). Environmentally Endangered Species Programme (EEP) captive Amur tigers have retained 96% of the genetic diversity of the wild population.

[2] On 31 Dec.1996, 501 (238 male, 263 female) registered Amur tigers were held in 175 facilities worldwide, according to the 1997 International Tiger Studbook. In order to maintain the quality of the stock, by Dec 2001 the total was reduced to 423 in 199 managed zoo populations, though there are many more unregistered animals in private hands. In the EEP (covering Europe and the former Soviet Union) on that date there were 215 (101 male, 114 female) Amur tigers in 87 institutions; the North American Species Survival Programme (SSP) had 153 (81 males, 72 females) in 57 institutions; the Japanese programme had 55 tigers (Christie 2000, 2001).

[3] Pocock (1929); Mazák (1967, 1979).

[4] Christie (1995a). Sankhala (1978, p.206) also felt that releasing captive tigers in India is unsatisfactory.

[5] Tiger breeding/reinforcement programmes have not always been carefully prepared. Ostensibly to help rescue Amur tigers from extinction in China, the Hengdaohezi Centre for Breeding Felidae was established in 1986 at a failed fur farm in the Wandashan. Later a second centre was established - the Northeast-China Tiger Park at Harbin. Concerns were expressed that the centres were being used as a cover for the trade in tiger derivatives as well as being a tourist spectacle, rather than serving any conservation purpose. Following international pressure the Chinese government agreed in 1993-94 to implement a ban on the trade in tiger parts, though this has not always been enforced.

By 1996 some 70 Amur tigers had been reared at the two centres. The tigers are quite inappropriately fed live cattle and chickens, supposedly to prepare them for release into the wild while at the same time providing a grisly show for visitors. Peter Jackson of the IUCN has pointed out that such released tigers - with little expertise in capturing wild prey - would associate food with people and vehicles, and would constitute a danger to both humans and domestic stock (Jackson 1996). These centres are of dubious conservation value. Some of the animals may not be pure bred Amur tigers and so would not be suitable for re-introduction into the wild, or reinforcement of existing wild populations. In any case little natural prey survives in the forests of Northeast China, and the ungulate populations would first need to be restored.

[6] Christie (1998). [7] Miquelle *et al* (2001).

[8] Tiger Response Team: see chapter 5, note 13. This particular tiger was donated to the Royal Museum, Edinburgh, and featured in the 2004 exhibition 'Cats' in the Museum.

[9] Miquelle *et al* (2001); Christie (2002). [10] Phoenix Fund Report (2002).

[11] Stamotyuk 1998. There were six zapovedniks in Primorski Krai in 1998, and the areas of Sikhote-Alin and Lazovski zapovedniks have been expanded. Borisovskoye Plateau Zakaznik has been added to nine others. There is a regional nature park, Khasanski, and work is in progress to create three national parks and two nature parks. Since 1992, plans for the conservation of biological diversity have been developed, based on the Long Term Programme for Wildlife Conservation and Rational Resource Use, in Primore, which began in 1992.

[12] 'A Gift from Primorsky Krai', *Russian Conservation News* no.30, Summer 2002, p.35.

[13] Miquelle *et al.* (1995a).

[14] Conservation in China is discussed in more detail in chapter 9.

[15] Cockburn (2000). [16] Inspection Tiger: see chapter 5, note 13.

[17] Matyushkin (1996). [18] Stamotyuk (1998).

[19] Deutsche-Press-Argentur, 27 Feb.2001. The provincial authorities were reported to have allocated 2.9 million roubles ($100,000) to monitor and protect the estimated 70 tigers in the Region.

[20] AP Worldstream, 23 Jan. 2001. [21] Pikunov *et al* (2002).

[22] Steve Galster, one of the Directors of WildAid (formerly known as the Investigative Network, set up to combat wildlife crime) played an important part in the organisation of anti-poaching teams in the Russian Far East in the early 1990s.

[23] Christie (1999).

[24] There is no evidence for the suggestion that poachers have had access to the equipment to enable them to pick up the signals to pursue the tigers.

[25] 21st Century Tiger 2001 Report.

[26] Ussuriyski Zapovednik is in the hills about 80 km north-east of Vladivostok.

[27] 20th December 2002 ITAR-TASS.

REFERENCES

Note: works are in English or Russian except where stated.

anon. *Oryx* 1998. Russian forest leased to Malaysian loggers, 32 (3), p.180; quoting *Taiga News*, issue 23, Jan.1998, 1.

Abramov V.K. 1970. *Areal i chislennost' tigra na Dal'nem Vostoke* (Range and Numbers of Amur Tiger in the Russian Far East), *Trudy 9 Mezhdunarodnogo Kongressa Biologov-okhotovedov, Moscow, September 1969.*

Abramov V.K. & Pikunov D.G. 1976. Redkiye vidy khishchnikh zverey yuga Dal'nego Vostoka SSSR (Rare Carnivores in the Southern Part of the Soviet Far East), in *Redkiye mlekopitayushchiye fauny SSSR* (Rare Mammals of the USSR Fauna), Moscow: Nauka. Pages 77-86.

Abramov V.K., Pikunov D.G., Bazylnikov V.I. & Sablina T.B. 1978. Oekologische Aspekte des Lebens der Tiger im Winter im Gebiet von Primorje (Ecological Aspects of the Winter Distribution of Tigers in Primor'e), *1. Internatio-nales Tigersymposium,* Zoologischer Garten Leipzig. In German and Russian.

Alexander, R.McNeill. 2003. Letter to *New Scientist*, 16 August, p.85.

Aramilev I. 1961. *Beyond the Ural Mountains,* London: George Allen & Unwin. Translated from Russian.

Arseniev V.K. 1939. *Dersu the Trapper,* London: Secker & Warburg.

Arsen'ev V.K. 1949. *Skvoz' taygu* (Through the Taiga), Moscow: Gos. izd. geograficheskoy lit-y.

Arsen'ev V.K. 1986. *Po Ussuriyskomu krayu* (Through the Ussuri Region), Vladivostok: Dal'nevostochnoye knizhnoye izd-o.

Arsen'ev V.K. 1994. *Dersu Uzala*, Tula: Izd.Peresvet II.

Astafiev A.A. 2001. Sikhote-Alinsky Zapovednik: Being an Advisor in the Russian Far East. *Russian Conservation News,* no. 27, pp.30-31. Washington DC.

Attenborough D. 2002. *The Life of Mammals,* London: BBC Books.

Bannikov A.G. 1974. *Po zapovednikam Sovetskogo Soyuza* (Through the Zapovedniks of the Soviet Union), Moscow: Mysl. 2nd edition.

Baykov N.A. 1915. *V gorakh i lesakh Man'chzhurii* (In the Mountains and Forests of Manchuria), Petrograd.

Baykov N.A. 1925. *Manch'zhurskiy tigr* (The Manchurian Tiger), Harbin: Obshchestvo izucheniya Man'chzhurskogo kraya.

Baykov N.A. 1927. Tigry na Dal'nem Vostoka (Tigers in the Far East), *Okhotnik* (Hunter), no.8, pp.23-24.

Baykov N.A. 1936. *Big Game Hunting in Manchuria* (English translation), London: Hutchinson.

Baykov N.A. 1939. *Les Bêtes Sauvages de la Mandchourie* (Wild Animals of Manchuria), Paris: Payot.

Baykov N.A. 1952. *Le Grand Van, la vie d'un tigre de Mandchourie* (The Great Van, the Life of a Manchurian Tiger), Paris: Payot.

Beltran J. (ed.) 2000. *Indigenous and Traditional Peoples and Protected Areas.* Gland, Switzerland & Cambridge UK: IUCN/WWF.

Benns M. 1998. Fearful Symmetry, *The Guardian Weekend,* London, 18 April.

Berg L.S. 1950. *Natural Regions of the USSR,* London: Macmillan.

Borisov, A.A. 1965. *Climates of the USSR*, Edinburgh & London: Oliver & Boyd. Translation of *Klimaty SSSR*, 2nd edition, Moscow, 1959.

Bragin A.P. 1986 preprint, 1989a. *Population Characteristics and Social-Spatial Patterns of the Tiger on the Sikhote-Alin' Eastern Macroslope.* Academy of Sciences of the USSR Pacific Institute of Geography, Vladivostok.

Bragin A. & Gapanov V. 1989. Problemy amurskogo tigra (Problems of the Amur Tiger), *Okhota i okhotnich'e khozyaystvo* (Hunting & Hunting Economy), no.10, pp.12-15.

Bromley G. 1964. *Ussuriyskiy kaban* (The Ussurian Wild Pig), Moscow: Nauka.

Cheng Jizhen 1999. Status and Management of Bears in Heilongjiang, China; in Servheen *et al.*, pp.123-5.

Chestin I. & Yudin V. 1999. Status and Management of the Asiatic Black Bear in Russia, in Servheen *et al.*, pp.211-13.

Christie S. 1995a. *European Studbook for Tigers*, no.1, Zoological Society of London.

Christie S. 1995b *Proceedings of the Special Meeting of the Amur Tiger EEP,* Moscow, 2-3 November 1995.

Christie S. 1996. *European Studbook for Tigers*, no.2.

Christie S. 1997. *European Studbook for Tigers*, no.3.

Christie S. 1998. *European Studbook for Tigers*, no.4.

Christie S. 1999. *European and Australasian Studbook for Tigers,* no.5.

Christie S. 2000. *European and Australasian Studbook for Tigers*, no.6. Also incorporates 2001.

Christie S. 2002. Translocating Tigers, in *Lifewatch,* Autumn/Winter, Zoological Society of London.

Cockburn P. 2000. Putin Puts Future Of Siberian Tiger at Risk, *The Independent,* London, 5 July 2000.

Court, W.E. 2000. *Ginseng: the genus Panax*, Australia &c: harwood academic.

Cracraft J, Feinstein J., Vaughn J., & Helm-Bychowski K. 1998. Sorting out Tigers *(Panthera Tigris)*: Mitochondrial Sequences, Nuclear Inserts, Systematics, and Conservation Genetics, *Animal Conservation,* 1: 139-50.

Danilkin A. 1996. *Behavioural Ecology of Siberian and European Roe Deer,* London: Chapman Hall.

Dezhkin, V.V. 1989 *V mire zapovednoy prirody* (In the World of Zapovednik Nature), Moscow: Sovetskaya Rossiya.

Deutz A., Catlin D., Laletin A., Teplyakov V. & Moshkalo V. 1999. *The Future of Forest Conservation in Russia,* Montreal & Moscow: IUCN Temperate and Boreal Forest Programme.

Digby B. 1928. *Tigers, Gold and Witchdoctors,* London: John Lane, the Bodley Head.

Dinerstein E. *et al.*, 1997. *A Framework for Identifying High Priority Areas and Action for the Conservation of Tigers in the Wild*, part 1, Washington DC: WWF/WCS.

Dinets V. & Rotshil'd E. 1996. *Zveri* (Animals), Moscow: ABF.

Dunishenko Yu. M. 1985. K voprosu okhrany tigra v Khabarovskom krae (The Problem of Tiger Conservation in Khabarovski Kray), in *Izucheniye i okhrana redkikh i izchezayushchikh vidov zhivotnikh fauny SSSR* (Study & Protection of Rare and Disappearing Animal Species of the USSR Fauna), Moscow: Nauka.

Pages 62-65.

Fisher S. 1996. *Wild Times,* Tusk Force, ed. 2, December.

Flerov K.K. 1952. *Musk Deer and Deer, Fauna of the USSR,* vol.1, no.2, Moscow: USSR Academy of Sciences. Translated published by Jerusalem: Israel Programme for Scientific Translations, 1960.

Flint V.E., Boehme R.L., Kostin Y.V. & Kuznetsov A.A. 1989. *A Field Guide to Birds of the USSR,* Princeton, NJ: Princeton University Press.

Formozov A.N. 1948. Pamyati L.G.Kaplanova (To the Memory of L.G.Kaplanov), in Kaplanov L.G. (1948). Reprinted 2000 in *Okhotnich'i prostory,* 4th book, (26), pp.164-186.

Forsyth J. 1992. *A History of the Peoples of Siberia,* Cambridge: Camb.Univ.Press.

Galster S. & Caldwell G.B. 1996. *Russia's Final Roar,* Washington DC: Investigative Network Report.

Galster S. & Eliot K.V. 1999. Roaring Back: Anti-Poaching Strategies for the Russian Far East and the Comeback of the Amur Tiger, in Seidensticker *et al.*.

Geist V. 1987. Bergmann's Rule is Invalid, *Canadian Journal of Zoology,* no. 65, pp.1035-8.

Geptner, see Heptner.

Gibson G. 1997. *Tigers,* Rusper, West Sussex: Care for the Wild International.

Goodrich J. 2003. *Siberian Tiger Project. Six Monthly Report to World Conservation Society 2003.*

Goodrich J., Kerley L., Miquelle D. & Smirnov E. 2002. Amur Tiger Conservation, Roads and Human Disturbance in the Russian Far East, *Cat News,* no.36.

Goulding M. 2003. *Wild Boar in Britain,* Whittet Books.

Gromov E.I. & Matyushkin E.N. 1974. K analizu konkurentnikh otnosheniy tigra i volka v Sikhote-Aline (Analysis of the Competitive Relations Between Tigers and Wolves in the Sikhote-Alin), *Biologicheskiye Nauki* (Biological Sciences), no.2, pp.20-25.

Guggisberg C.A.W. 1975. *Wild Cats of the World,* London: David & Charles.

Harper G.H. & Peterken G.F. 2003. Zapovedniks - scientific nature reserves in Russia. *British Wildlife,* 15(2): 85-90.

Hemmer H. 1981. Die Evolution der Pantherkatzen Modell zur Überprufung der Brauchbarkeit der Hennigschen Prinzipien der Phylogenetischen Systematik fur Wirbeltierpalaeontologische Studien. *Palaeontologische Zeitschrift,* 55: 109-116.

Hemmer H. 1987. The Phylogeny of the Tiger, in Tilson & Seal, pp.28-35.

Heptner V.G., Nasimovich A.A. & Bannikov A.G. 1989. *Mammals of the Soviet Union,* vol.1, Ungulates, Leiden: E. J. Brill. Translated from the 1961 Russian edition.

Heptner V.G. & Sludskii A.A. 1992. *Mammals of the Soviet Union,* vol.2, part 2 Carnivora (Hyaenas & Cats), Leiden: E.J. Brill. Translated from the 1972 Russian edition.

Herrington S. 1987. Subspecies and the Conservation of *Panthera tigris.* Preserving Genetic Heterogeneity; in Tilson & Seal 1987, pp 51-60.

Iribarren I.B. 2001. The History of the Biosphere Reserve Concept, *Russian Conservation News,* no.27, pp.6-8.

Isaev A. 2002. Russia's Changing Forest World, *Russian Conservation News,* no 28.

Jackson P. 1990. *Endangered Species: Tigers,* London: Apple Press.

Jackson P. & Kemf E. 1994. *Wanted Alive! Tigers in the Wild.* A WWF Species Status Report. Gland, Switzerland: WWF.

Jackson P. 1996. Concern About Chinese Tiger Programme, *Cat News*, no.25, pp.13-14.

James H.E.M. 1888. *The Long White Mountain, or a Journey in Manchuria,* London.

Kaplanov L.G. 1948. *Tigr, izyubr', los'* (Tiger, Manchurian Red Deer, Elk), Moscow: Izd. *Mosk. obshchestva ispytateley prirody.*

Kemf E. (ed.) 1993. *Indigenous Peoples and Protected Areas, the Law of Mother Earth.* London: Earthscan Publications.

Khramtsov V.S., 1995. O povedenii tigra pri vstreche c chelovekom (Tigers' Behaviour in Encounters with Man), *Ecologiya no*.3, pp.252-254.

Khramtsov V.S. & Zhivotchenko V.I. 1983. Tigers' Attacks on Bears in Lazovskiy Zapovednik, *Redkiye vidy mlekopitayushchikh SSSR i ikh okhrana*, Moscow: USSR Academy of Sciences. Pages 88-89.

Kirby E.S. 1971. *The Soviet Far East*, London: Macmillan.

Kitchener A.C. 1991. *The Natural History of the Wild Cats,* London: Christopher Helm, A. and C. Black.

Kitchener A.C. 1997. The Role of Museums and Zoos in Conservation Biology, *International Zoo Yearbook,* vol.35, pp.325-336. Zoological Society of London.

Kitchener A.C. 1999. Tiger Distribution, Phenotypic Variation and Conservation Issues, in Seidensticker *et al* 1999.

Kitchener A.C. & Dugmore A.J. 2000. Biogeographical change in the tiger *Panthera tigris*, *Animal Conservation*, 3: 113-124.

Knystautas A. 1987. *The Natural History of the USSR*, London: Century Press.

Kolarz W. 1954. *The Peoples of the Soviet Far East*, New York: Philip.

Komarov V.L. & Ilin M.M. 1934. *Flora SSSR* (Flora of the USSR), volume 1, Leningrad: Akademiya nauk SSSR.

Korkishko V.G & Pikunov D. 1994. *The Population Numbers of the Far Eastern Leopard,* Gland, Switzerland: IUCN.

Kostomarov S.V. & Matyushkin E.N. 1997. Perviye shagi Botchinskogo zapovednika (The First Steps of Botchinskiy Zapovednik), *Okhota i okhotnich'e khozyaystvo* (Hunting and Hunting Economy), no 2, pp.4-7.

Kucherenko S.P. 1970. Amurskiy tigr (sovremennoe rasprostraneniye i chislennost') (Amur Tiger (Present Distribution and Numbers)), *Okhota i okhotnich'e khozyaystvo,* (Hunting and Hunting Economy), no 2, pp.20-23.

Kucherenko S.P. 1983. Bylaya chislennost' Amurskogo tigra (Former Numbers of Amur Tiger), *Redkie vidy mlekopitayuschikh SSSR i ikh okhrana* (Rare Mammals of the USSR and their Protection), pp.122-3. Moscow: Nauka.

Kuchler A.W. 1948. A New Vegetation Map of Manchuria, *Ecology,* 24.

Laptev A. 2003. New methods for a new era of conservation in Lazovsky Zapovednik. *Russian Conservation News*, no33, pp.13-15.

Levin M.G. & Potapov L.P. (eds.) 1964. *The Peoples of Siberia,* University of Chicago Press. Chapter on Udege, pp.737 49. Translated from 1956 Russian edition.

Li Wenhua & Zhao Xianying 1989. *China's Nature Reserves,* Beijing: Foreign Language Press.

Loukashkin A.S. 1938. The Manchurian Tiger, *China Journal,* vol.28, no.3, pp.127-133.

Loukashkin A.S. & Jernakov V. 1934. The Fur Trade of Manchuria, *China Journal,* vol.21, no.5.

Lu Houji 1987. Habitat Availability and Prospects for Tigers in China; in Tilson & Seal, pp.71-74.

MacKinnon J. 1996. *Wild China,* London: New Holland Publishers.

Ma Yiqing, Li Xiaomin & Sheng Helin 1997. Status and Conservation of Tigers in China, *Cat News,* no.26, pp.6-7.

Maak R. 1861. *Puteshestviye po doline Ussuri* (Journey through the Ussuri Valley), St. Petersburg.

Mainka S.A. 1997. Tiger Progress? The Response to CITES Resolution Conf. 9.13, Cambridge UK: TRAFFIC International.

Malyshev A.A. 1966. *Povest' o taezhnom sledopyte* (Story of a Taiga Tracker), Vladivostok. New edition 1971.

Massey Stewart J. 1992. *The Nature of Russia,* London: Boxtree.

Matyushkin E.N. 1973. Tigr i chelovek: problemy sosedstva (Tiger and Man - the Problem of Living Together), *Priroda.* no.12, pp.82-88.

Matyushkin E.N. 1974. Sledi Amurskogo tigra (Tracks of the Amur Tiger), *Okhota i okhotnich'e khozyaistvo* (Hunting & Hunting Economy), no.5, pp.12-17.

Matyushkin E.N. 1977. Vybor puti i osvoyeniye territorii Amurskim tigrom (Routes and Territory Used by the Amur Tiger, According to Winter Tracking Data), *Povedeniye mlekopitayushchikh* (Behaviour of Mammals), USSR Academy of Sciences, Moscow: Nauka. Pages 146-178.

Matyushkin E.N. 1978. Besonderheiten zur Verbreitung und Oekologie des Amurtigers als einer Geographischen Form der Art, die sich an extreme Unweltbedingungen angepasst hat (Characteristic Features of the Distribution and Ecology of the Amur Tiger as a Geographical Form of a Species Adapted to Extreme Environmental Conditions) *1. Internationales Tigersymposium,* 11-12 October 1978, Leipzig Zoological Gardens. In German and Russian.

Matyushkin E.N. 1992. Tigra i izubr na primorskikh sklonakh srednego Sikhote-Alinya (Tigers and Manchurian Red Deer on the Maritime Slopes of the Sikhote-Alin), *Bull. Mosk. obshchestva ispytateley prirody,* otd. biol., 97 (1): 3-20.

Matyushkin E.N. 1993 Mirooshchushcheniye naturalista (World View of a Naturalist) (pp.2-6), Pis'ma L.G.Kaplanova (Letters of L.G.Kaplanov) (pp.7-17), *Bull. Mosk. obshchestva ispytateley prirody,* 1993 (3): 2-17.

Matyushkin E.N. (ed.) 1996. *Strategy for Conservation of the Amur Tiger,* Moscow & Vladivostok: USAID/WWF.

Matyushkin E.N. 1998. *The Amur Tiger in Russia: An Annotated Bibliography 1925-1997,* Moscow: WWF/USAID.

Matyushkin E.N. *et al.* 1996. *Numbers, Distribution and Habitat Status of the Amur Tiger in the Russian Far East,* Moscow: USAID/WWF.

Matyushkin E.N. & Formozov A.N. 1998. The World through a Naturalist's Eyes, *Russian Conservation News*, no.15: 8-9. (About Kaplanov.)

Matyushkin E.N. & Krever V.G. (eds.) 1999. *Strategy for Conservation of the Far Eastern Leopard in Russia,* Moscow & Vladivostok: WWF /USAID.

Matyushkin E.N., Zhivotchenko V.I. & Smirnov E.N. 1978. *Tigers in the Past, Present and Future in the USSR.* Moscow: Main Department of Nature Conservation, Reserves, Forest & Game Management. Pages 104-125.

Matyushkin E.N., Zhivotchenko V. I. & Smirnov E.N. 1980. *The Amur Tiger in the USSR,* Gland, Switzerland: IUCN.

Mazak V. 1967. Notes on the Siberian Long Haired Tiger, *Mammalia,* 31 (4): 537-573.

Mazak V. 1979. *Der Tiger, Panthera Tigris,* Die Neue Brehm-Bucherei,Wittenberg-Lutherstadt. In German.

Mazak V. 1981. Panthera tigris, *Mammalian Species,* 152: 1-8.

Melenina I. 2001. Transcript of Speech to Tumen Project Council Meeting, Sokch'o, Republic of Korea. 29-30 Jan.2001.

Mesitov A.M. 1994, in Arsen'ev 1994.

Mills J. & Jackson P. 1994. *Killed for a Cure; A Review of the World-Wide Trade in Tiger Bone.* Cambridge UK: TRAFFIC International.

Miquelle D.G, 1998a. Sharing the Spotlight, *Zoogoer* (Friends of the National Zoo), Washington, D.C., September, 27 (5).

Miquelle D.G. 1998b. Tigers and Leopards in Jilin Province China, *Cat News,* no.28, Spring, pp.5-6. (Bougy, Switzerland: IUCN Cat Specialist Group.)

Miquelle D.G. 2000. Counting Tigers in the Russian Far East: 'How Many Are There?' versus 'Is there a Change', *Russian Conservation News*, no.23, Summer 2000, pp.26-27.

Miquelle D.G. *et al.* 1999a. A Habitat Protection Plan for the Amur Tiger: Developing Political and Ecological Criteria for a Viable Land-Use Plan; in Seidensticker *et al.* 1999.

Miquelle D.G. *et al.* 1999b. Hierarchical Spatial Analysis Of Amur Tiger Relationships to Habitat and Prey, in Seidensticker *et al.* (1999).

Miquelle D.G., Arzhanova T.D. & Solkin V.A. (eds.) 1996b. *A Recovery Plan for Conservation of the Far Eastern Leopard: Results of an International Conference held in Vladivostok*. Final Report to USAID.

Miquelle D.G., Goodrich J. & Christie S. 2001. *Rehabilitation and Translocation of Two Adult Female Amur Tigers,* New York: Wildlife Conservation Society.

Miquelle D.G., Quigley H.B. & Hornocker M.G. 1995a. *A Habitat Protection Plan for Amur Tiger Conservation*, Moscow, Idaho: Hornocker Wildlife Institute.

Miquelle D.G., Quigley H.B., Hornocker M., Smirnov E.N. & Nikolaev I.G. 1995b. Past and Present Status, Present Threats, and Conservation of the Siberian (Amur) Tiger in the Wild, in Christie S. 1995b. Pages 18-30.

Miquelle D.G., Smirnov E.N., Quigley H.B., Hornocker M. & Nikolaev I.G. 1996. Food Habits of the Amur tigers in Sikhote-Alin Reserve and the Russian Far East: Implications for Conservation, *Journal of Wildlife Research,* 1(2): 138-47.

Mountfort, G. 1973. *Tigers*, Newton Abbot: David & Charles.

Mowat, F. 1952. *People of the Deer*, London: Michael Joseph.

Mueller P. 1995. The International Tiger Studbook as a Basis for Regional and Global Breeding Programmes, in Christie, S. 1995b. Pages 7-11.

Myslenkov A.I. & Voloshina I.V. 1989. *Ecologiya i povedeniye Amurskogo gorala* (Ecology and Behaviour of the Amur Goral), Moscow: Nauka.

Novikov G.A. 1962. *Carnivorous Mammals of the Fauna of the USSR*, Jerusalem: Israel Programme for Scientific Translations. Pages 78-85.

Nowell K. & Jackson P. 1996. *Wild Cats, Status Survey and Conservation Action Plan,* Gland, Switzerland: IUCN.

Nowell K. 2000. *Far from a Cure: the Tiger trade Revisited,* Cambridge: TRAFFIC International.

O'Brien S.J. *et al.* 1987. Setting the Molecular Clock in Felidae: the Great Cats, *Panthera*; in Tilson & Seal, pp.36-50.

Ognev S.I. 1935. *Mammals of the USSR & Adjacent Countries,* Vol.3, Carnivora, Jerusalem: Israel Programme for Scientific Translations. Translated from 1962 Russian edition.

Ortolani A. & Caro T.M. 1996. The Adaptive Significance of Colour Patterns in Carnivores, in Gittleman (ed.) *Carnivore Behaviour, Ecology and Evolution,* Ithaca: Cornell University Press. Pages 132-188.

Oryel L. 2001. Primorski Krai: Good Hunting Does Not Mean Successful Marketing, *Zolotoi Rog* (Golden Horn), no.15, Vladivostok.

Oshmarin P.G. & Pikunov D.G. 1990. *Sledi v prirode* (Tracks in the Wild), Moscow: Nauka.

Payne-Gallwey R. 1995. *The Crossbow*, London: Holland Press. 1st edition 1903.

Perry R. 1964. *The World of the Tiger,* London: Cassell.

Peyton B., Servheen C., & Herroro S. 1999. An Overview of Bear Conservation Planning and Implementation, in *Bears: Status and Conservation Action Plan*, Gland, Switzerland & Cambridge, UK: IUCN.

Pikunov D.G. 1988a. Eating Habits of the Amur Tiger in the Wild, *Proceedings of the Fifth Conference on Breeding Endangered Species in Captivity,* Cincinnati. Pages 185-190.

Pikunov D.G. 1988b. Amur Tiger: Present Situation and Perspectives for its Preservation in the Soviet Far East. *Proceedings of the Fifth Conference on Breeding Endangered Species in Captivity,* Cincinnati. Pages 175-185.

Pikunov D.G. 1990. *Chislennost' tigrov na Dal'nem Vostoke* (The Numbers of Tigers in the [Soviet] Far East), Moscow: All-Soviet Theriological Society, USSR Academy of Sciences.

Pikunov D.G. 1992. Strategy of Conservation of Rare Animal Species in View of the Economic Development of the Far East. Ecol. *Aspecty razvitiya proizvodit. sil. Dal'nego Vostoka,* Moscow: Nauka. Pages 82-92.

Pikunov D.G. 1994. Proposals for Conserving the Amur Tiger in the Russian Far East, *Cat News,* no.20, pp.4-6.

Pikunov D.G. *et al.* 1999. *Strategy for Conservation of the Far Eastern Leopard in Russia,* Moscow & Vladivostok: WWF.

Pikunov D.G., Bazylnikov V. I. & Bromley G.F. 1978. *Beutesuche und Beutefang des Amurtigers* (Search and Capture of Prey of the Amur Tiger), *1 Internationales Tigersymposium*, Leipzig, pp.13-23. In German and Russian.

Pikunov D.G. & Korkishko V.G. 1992. *Leopard Dal'nego Vostoka* (The Far Eastern Leopard), Moscow: Nauka.

Pikunov D.G. & Miquelle D. 1998. Tigers in North Korea, *Cat News*, no.28.

Pikunov D.G., Miquelle D. & Fomenko P. 2002. Deep Snow Threatens Far Eastern Leopards, *Cat News*, no.36.

Pocock R. 1929. Tigers, *Journal of the Bombay Natural History Society*, 33: 505-541.

Pocock R. 1931. The Panthers and Ounces of Asia, *Journal of the Bombay Natural History Society*, 34: 64-82.

Pocock R. 1935. The Black and Brown Bears of Europe and Asia, *Journal of the Bombay Natural History Society*, 35: 771-823.

Poddubnaya N. & Kovalev V. 1993. Tigr v Ussuriyskom zapovednike (Tigers in the Ussuriskiy Zapovednik: Status and Prospects), *Bull. Mosk. obshchestva ispytateley prirody* (Bulletin of the Moscow Society of Naturalists), 98 (3): 54-61.

Potapov R. 1976. *Tigrovaya Balka. Nevedomyy Pamir* (Tigrovaya Balka. Unknown Pamir), Moscow: Mysl'.

Pryde R.R. 1972. *Conservation in the Soviet Union,* Cambridge: Camb.Univ.Press. Page 69.

Pryde R.R. & Mote V. 1990. Environmental Constrains and Biosphere Protection in the Soviet Far East; in Rodgers (1990).

Prynn D. 2002. The Amur leopard is in peril, *Marwell Zoo News,* no.112 (autumn 2002), 10-11.

Prynn D. 2003. The Caspian Tiger: a lesson from history, *Marwell Zoo News,* no.116 (summer 2003), 10-11.

Rodgers A. (ed.) 1990. *The Soviet Far East, Geographical Perspectives on Development,* London: Routledge.

Salkina G.P. 1993. Sovremennoye sostoyaniye populyatsii tigra na yugo-vostoke Sikhote-Alinya (Current Status of the Population of the Amur Tiger in the South-East of the Sikhote-Alin), *Bull. Mosk. obshchestva ispytateley prirody* (Bulletin of the Moscow Society of Naturalists, Biological Series), 98: 45-53.

Salkina G.P. 1994. *Tigr v Lazovskom zapovednike* (The Tiger in Lazovski Zapovednik), Vladivostok: Russian Academy of Sciences, Far Eastern Branch.

Salkina G.P. 2000. In the Russian Far East: Fewer Ungulates Ominous Sign for Siberian Tiger, *Russian Conservation News,* no.23, Summer 2000.

Sankhala K. 1978. *Tiger! The Story of the Indian Tiger,* London: Collins.

Seidensticker J. 1996. *Tigers,* Grantown-on Spey: Colin Baxter.

Seidensticker J., Christie S. & Jackson P. (eds.) 1999. *Riding the Tiger: Tiger Conservation in Human-dominated Landscapes,* Cambridge: Camb.Univ.Press.

Servheen C., Herroro S. & Peyton B. 1999. *Bears: Status and Conservation Action Plan,* Gland, Switzerland & Cambridge, UK: IUCN.

Shackleton D.M. 1997. *Wild Sheep and Goats and Their Relatives: Status Survey and Action Plan for Caprinae.* Gland, Switzerland & Cambridge, UK: IUCN.

Shafer E.H. 1968. Hunting Parks and Animal Enclosures in Ancient China, *Journal of Economic and Social History of the Orient,* 2: 318-343.

Shibnev Yu. 1989. The Deer Hunter, *BBC Wildlife,* August 1989, vol.7, no.8.

Shtil'mark F. 2002. *Na sluzhbe prirode i nauke* (In the Service of Nature and Science), Moscow: Logata.

Shtil'mark F. 2003. *History of the Russian Zapovedniks 1895-1995,* Edinburgh: Russian Nature Press. Translated from 1996 Russian edition.

Silantiev A.A. 1898. *Obzor promyslovykh okhot v Rossii* (Review of Hunting in Russia), St. Petersburg.

Simplich F. 1929. Manchuria, Promised Land of Asia, *National Geographic Magazine,* vol.LVI, no.4, October 1929, pp.379-428.

Smirin V.M. & Smirin Yu.M. 1999. *Animals in Nature,* Edinburgh: Russian Nature Press. Translation of *Zveri v prirode,* Moscow: Moscow University Press, 1991 (2nd edition Moscow: Armada, 2001).

Smirnov E.N. 1993. Gruppirovka tigrov Sikhote-Alinskogo zapovednika i sopredel'noy territorii v 80-e gody (The Tiger Population of the Sikhote-Alin' Zapovednik and Adjacent Territory in the 1980s), *Bull. Mosk. obshchestva ispytateley prirody* (Bulletin of the Moscow Society of Naturalists), 98 (3): 37-44.

Smirnov E.N. 2000. *Vstrechi c tigrom* (Encounters with Tigers), Vladivostok: Dalnauka.

Smirnov E.N. & Miquelle D.C. 1999. Population Dynamics of the Amur Tiger in Sikhote-Alin Zapovednik, Russia; in Seidensticker *et al.* 1999.

Sowerby A. de C. 1923. *The Naturalist in Manchuria,* vol.2, Tientsin Press.

Sowerby A. de C. 1933. The Tiger in China, *The China Journal* (Shanghai), 18 (2): 94-101.

Sowerby, A. de C. 1934. Game in the Maritime Province, *The China Journal* (Shanghai), 21(1): 40-44.

Sowerby A. de C., 1937. The Mammals of China, Mongolia, Eastern Tibet, and Manchuria Requiring Protection, *The China Journal,* vol.27, no.5, May 1937.

Spitsin V.V., Romanov P.N., Popov S.V. & Smirnov E.N. 1987. The Amur Tiger (*Panthera tigris altaica* Temminck 1844) in the USSR: Status in the Wild and in Captivity; in Tilson & Seal, pp.64-70.

Stamotyuk E. 1998. Conservation of the Amur Tiger in Primorski Krai, *Cat News,* no.29, pp.2-4.

Stephan J.J. 1994. *The Russian Far East: A History,* Stanford University Press.

Strand H. 1995. The Russian Far East; in Pryde P. R. *Environmental Resources and Constraints in the Former Soviet Republics,* Boulder, San Francisco, & Oxford: Westview Press. Pages 79-94.

Stroganov S.U. 1962. *Carnivorous Mammals of Siberia,* Moscow: USSR Academy of Sciences. Translated 1969, Jerusalem: Israel Programme for Scientific Translations.

Sun Baogang, Endi Zhang & Miquelle D.G. 1999. Siberian Tigers on Brink of Extinction in China, *Cat News,* no.31, p. 2.

Sunquist. M., Karanth U.K. & Sunquist F. 1999. Ecology, Behaviour and Resilience of the Tiger and its Conservation Needs; in Seidensticker J. *et al.*, 1999.

Suslov S.P. 1961. *Physical Geography of Asiatic Russia,* San Francisco & London: Freeman. Translation of 1947 Russian edition.

Sysoyev V. 1960. *Okhota v Dal'nevostochnoy taige* (Hunting in the Far Eastern Taiga), Khabarovsk: Khabarovsk Book Publishers.

Taksami Ch. M. 1982. *Kraski zemli Dersu* (Colours of the Land of Dersu), Khabarovsk: Khabarovsk Book Publishers.

Tan Banjie 1987. Status and Problems of Captive Tigers in China; in Tilson & Seal, pp.134-48.

The Times Special Correspondent, 1976, Animals Protected as Natural Resource, in China, a Special Report, *The Times*, London, 28th September, p.13.

Thiel S. 1957. *The Soviet Far East.* London: Methuen.

Thubron C. 2000. *In Siberia,* London: Penguin Books.

Tilson R.L. & Seal U.S. (eds.) 1987. *Tigers of the World: the Biology, Biopolitics, Management and Conservation of an Endangered Species,* Park Ridge, New Jersey: Noyes Publications.

Tkachenko K.N. 1996. Tigr v Bolshekhekhtsirskom zapovednike (Tigers in the Bolshekhekhtsir Zapovednik), *Zoologicheskiy zhurnal* (Zoological Journal), 75 (11): 1729-1736.

TRAFFIC International, 1994. Analysis of the Market for Tigers, Bears and Musk Deer in the Russian Far East. *TRAFFIC Bulletin,* 15 (1): 23-30.

Vakhreev G.I., Yudin V.G. & Nikolaev I.G. 1983. Svyaz morfologicheskikh osobennoctey tigra c vozrastom (Relations between Morphological Characteristics and Age in Tigers), in *Redkiye vidy mlekopitayushchikh SSSR i ikh okhrana* (Rare Mammals of the USSR and their Protection), Moscow: Nauka. Pages 88-89.

Vasilev N.G. & Matyushkin E.N. 1982. *Rastitel'nyy i zhivotnyy mir Sikhote-Alinskogo zapovednika* (Plant and Animal World of Sikhote-Alin Zapovednik), Moscow: Nauka.

Veselovsky Z. 1967. The Amur Tiger in the Wild and in Captivity, *International Zoo Yearbook,* vol.7, Zoological Society of London.

Voloshina I.V., Elsukov S.V. & Vdovin A.N. 1999. *Kadastr pozvonochnikh zhivotnykh Sikhote-Alinskogo zapovednika i severnogo Primor'ya* (Check list of Vertebrates of the Sikhote-Alin Zapovednik and Northern Primorye), Vladivostok: Dalnauka.

Wallace M.P. 2000. Retaining Natural Behaviour in Captivity for Re-introduction Programmes; in Gosling L.M. & Sutherland W.J. *Behaviour and Conservation.* Cambridge: Camb.Univ.Press.

Wang Chi-Wu 1961. *The Forests of China,* Harvard University.

Wemmer C. (ed.) 1998. *Deer: Status and Conservation Action Plan*, IUCN. Gland Switzerland, &Cambridge: IUCN.

Wenzel J. *et al. 1999.* Subspecies of Tigers: Molecular Assessment Using "Voucher Specimens" of Geographically Traceable Individuals; in Seidensticker, J. *et al.* 1999.

Williamson, The Rev. A. 1870. *Journeys in North China, Manchuria and Eastern Mongolia with some account of Corea*, London: Smith Elder, vol.2, p.68.

Wu Xianzhong 1994. (Population Size and Distribution Status of the Amur Tiger in Heilongjiang Province), *Chinese Wildlife,* no.3, 17-20. In Chinese.

Xiaochen Yu, Endi Zhang & Miquelle D.G. 2000. Monitoring the Tiger Population in Heilongjiang Province, China. *Cat News,* no.33, pp.3-4.

Xingjia Yang & Jinsong Jiang, 1996. Distribution, Numbers and Conservation of the North China Leopard in Jilin Province, China; in Miquelle *et al* 1996b.

Yudakov A.G. & Nikolayev I.G. 1973. Sostoyaniye populyatsii Amurskogo tigra v Primorskom krae (State of the Population of the Amur Tigers in Primorski Kray), *Zoologicheskiy zhurnal* (Zoological Journal), 52 (6): 909-919.

Yudakov A.G. & Nikolayev I.G. 1987. *Ecologiya Amurskogo tigra* (Ecology of the Amur Tiger), Moscow: Nauka.

Yudin V.G. 1992. *Volk Dal'nego Vostoka Rossii* (The Wolf of the Russian Far East), Blagoveshchensk: Russian Academy of Sciences, Far Eastern Branch.

Zabelina N.M., Isayeva-Petrova L.S. & Kuleshova L.V. 1998. *Zapovedniki i natsional'nyye parki Rossii* (Zapovedniks & National Parks of Russia), Moscow: Logata.

Zdansky O. 1924. Jungtertiäre Carnivoren, *Palaeontologia Sinica*, series C, 2(1): 1-145.

Zhao Ji, Zheng Guangmei, Wang Huadong, & Xu Jialin. 1990. *The Natural History of China*, London: Collins.

Zhidenkov, B. 1966. *Po tigrovomu sledu* (Tracking Tigers), Vladivostok: Dal'nevostochnoye knizhnoye izdatel'stvo.

Zhivotchenko V.I. 1981. O pitanii Amurskogo tigra (Food Habits of the Amur Tiger), *Khishchnyye mlekopitayushchiye* (Mammalian Predators), Moscow: Glavokhoty RSFSR. Pages 64-75.

Zhivotchenko V.I. 1985. Amurskiy tigr: problemy okhrany i upravleniya (The Amur Tiger, Problems of Conservation and Management), *Okhota i okhotnich'e khozyaystvo* (Hunting and Hunting Economy), no.6, pp.10-11.

Zhivotchenko V.I. *et al.* 1989. *Lazovskiy Zapovednik,* Moscow: Agropromizdat. Essay on the tiger, pp.102-113.

INDEX

Words appearing very frequently in the text, such as 'pig', 'tiger', 'Primore', 'Khabarovsk Region' and 'Russian Far East' are not included, or only selectively. In some extended page sequences, such as '110-18', the item may not be mentioned on all of the pages.

The Notes are indexed only sketchily, and some appendices not at all. For names of animals and plants, see Appendix 1; for a summary of the different tigers, see Appendix 2; for NGOs and projects, see Appendix 3.

Abbreviations: App(endix); L(ake); Mts - Mountains; n(ote); R(iver).

IN MEMORIAM

Evgeni Nikolayevich Matyushkin

11.viii.41 - 27.ix.03

A remarkable person has departed - one of the 'last of the Mohicans' among the great 'tribe' of Russian naturalists. He was a person of amazing gentleness, sensitivity, and genuine intelligence, combining openness to people with a intense inner life. His spirit was not coarsened by the demanding winter taiga journeys lasting many days, the interminable bivouacs beside the camp fire, or all the burdens that are borne on the shoulders of the field ecologist; on the contrary, it became even more open towards people and nature.

Zhenya Matyushkin was born in Saratov, where his family was evacuated early in the war, and from 1956 he was a member of the Young Biologists' Club at Moscow Zoo. In 1959 he entered the biogeography department of the Geography Faculty at Moscow State University. He graduated in 1964 and, with his wife Lyudmila Vladimirovna Kuleshova, found work in the Sikhote-Alin Zapovednik. Although he did not remain there long, the experience coloured the rest of his life and shaped both his scientific interests and nature-conservation ethics. Like a magnet the Sikhote-Alin attracted Evgeni Nikolayevich all his life, and he returned there many times to carry out field work. On the basis of observations made in the zapovednik, in 1970 he was awarded a PhD on 'Territorial Associations in the Central Sikhote-Alin, and the Origin of its Mammalian Fauna'. This work raised the fascinating question of how to classify the geographic and genetic elements of the fauna and analyse their inter-relations, and also introduced the 'faunule' concept as the elemental faunistic association. The subject interested Evgeni Nikolayevich throughout his life. There too, in the Russian Far East, he became intrigued by large-scale disjunct distributions and the reconstruction of their history.

It was in the Sikhote-Alin that Matyushkin first came across the large cats - a group on which he became a acknowledged expert. His first article on Amur tiger conservation appeared in 1966. During the 1990s he headed a team developing the 'Conservation Strategy for the Amur Tiger in Russia', and on his initiative a comprehensive survey of the tiger in Russia was carried out in the winter of 1995-96. Evgeni Nikolayevich was dedicated to the protection and study of the large cats until his very last breath. His Russian- and English-language bibliographic handbook *The Amur Tiger in Russia* opened up for 'Western' biologists the massive fund of knowledge about the Amur tiger that had been gathered by the sweat and blood of several generations of Russian naturalists. Matyushkin's last monograph - *The Lynx* (in the series 'Commercially

Valuable Wild Animals of Russia and Neighbouring States, and their Habitats') - was still being prepared for the press when he was already in hospital. He dictated the Conclusion to his wife and son, and managed to see copies of the published work, signing some for his friends as a keepsake.

Another of E.N's scientific interests was winter ecology. His field trips encompassed practically all the forested areas of northern Eurasia, and the 'zone of deep loose snow' was his favourite habitat. The love of winter ecology was yet another result of his demanding field work in the Sikhote-Alin. In this pursuit he rightly regarded himself as the pupil and successor of Aleksandr Formozov. A general review of the subject appeared in the issue of *Zoologicheski Zhurnal* (no.4, 2000) devoted to Formozov's memory - 'The use of signs and tracking in the study of large mammalian predators'. Altogether Matyushkin published six monographs and over a hundred papers. The camera played a special role in his publications: unique photographs give us an idea of the surroundings as seen through the eyes of the animals themselves, and at the same time present an unforgettable and comprehensive portrait of the landscapes that he described so exactly and, on occasion, poetically.

Evgeni Nikolayevich made a significant contribution to the work of the Academy of Sciences Zapovedniks Commission. He firmly believed that all means possible should be used to protect and support zapovednik science - the linchpin of the zapovednik system in Russia; it was science that the zapovedniks were created for and thanks to which they still exist.

During a quarter of a century Evgeni Nikolayevich was on the editorial board, and from 1981 deputy editor, of Russia's oldest scientific journal *Bulletin of the Moscow Society of Naturalists, Biological Section.* He saw it as his duty to help zapovednik scientists and other experts working in remote areas to publish their sometimes unique observations and results.

It is hard to overestimate what E.N.Matyushkin - so prematurely taken from us - did for Russian nature, for our understanding of it, and for the people who study it.

N.Formozov & E.Pavlova

DAVID PRYNN

David Prynn B.Sc. (Econ., London), M.A. (Sheffield), has been involved in Russian and Chinese studies since 1969. Before his retirement he was a Lecturer in History at Suffolk College, Ipswich, and an Associate Lecturer for the Open University. His research on open-air movements, particularly the Woodcraft Folk (published as a series of articles for the *Journal of Contemporary History*), brought together his interests in social history and nature.

An enthusiasm for the wild regions of the Russian Far East and Northeast China (Manchuria) and its wildlife, including the Amur tiger, was stimulated not only by the writings of explorers and naturalists such as Arsenev and Baykov, but also by tours to Northeast China in 1986, and the Russian Far East in 1992 and 2001. These included most of the zapovedniks containing Amur tigers. On the basis of these field study tours, and also visits to museums, zoological collections and libraries with material on the fauna and indigenous people of the Manchurian region, several articles have been published in *Oryx* and in *Marwell Zoo News*.

David has continued his research on the Amur tiger at the University of East Anglia, and was awarded an MPhil in July 2003. He is a Fellow of the Zoological Society of London and a Trustee of AMUR - an Anglo-Russian charity devoted to the conservation of the Amur tiger and leopard.

GEOFFREY HARPER

A naturalist from boyhood, Geoffrey acquired an interest in Russian while a botany student at Cambridge in the 1960s. After several part-time attempts to learn Russian over the decades, he at last made a proper job of it by training as a translator at Heriot-Watt University, Edinburgh, graduating in 1992. Meanwhile he has made numerous trips to the former Soviet Union, including the Baltic States; Kola Peninsula; the Altay, Putorana, Tyan Shan and Ural Mountains; Kamchatka; and the Central-Forest and Aksu-Jabagly zapovedniks. Seven months of 1991 were spent at Akademgorodok, Novosibirsk (winter was best; summer - too hot).

Geoffrey is currently employed as a botanist at the Royal Botanic Garden Edinburgh, and also runs a phenology project there to study the effects of climate change.

Cherga, Altay Mountains, 1991

RUSSIAN NATURE PRESS

For much of the 20th century the English- and Russian-speaking worlds of natural history, ecology and nature conservation worked largely in isolation. The language barrier led to some absurd situations - such as Jim Lovelock developing his Gaia theory while remaining largely unaware of Vladimir Vernadski's related work on the biosphere concept, published in 1926.

The language barrier, however, may not have been entirely a bad thing. For instance it contributed to the isolation of Russian culture and hence perhaps also to the development of the unique system of Russian scientific nature reserves (zapovedniks) - as described in Feliks Shtilmark's book.

Now the language barrier probably does more harm than good. The time has come to bring the two worlds together. While Russian-speaking ecologists and conservationists tend to be rather good at learning English and finding out what we are doing, we are not so adaptable - and so we are the losers. We should learn about the Russians' achievements, and we might benefit from their experience. It is time for naturalists, ecologists and conservationists all over the world to cooperate by sharing knowledge, experience, ideas and inspiration.

To encourage this process, RNP translates works from Russian that will help stimulate interest among non-specialists. It is also intending to translate specialist works (such as Shtilmark's book) that ought to be in English. The present book represents RNP's first original work. Read about our books in the following pages.

While we welcome sponsorship from a few generous donors, who help cover printing costs, we hope that many of you will support RNP's work by buying our books (they make excellent presents!) and telling your friends about them. Information and ordering details are available on the website or by writing to RNP at 19-3 Warriston Road, Edinburgh EH7 4HN, United Kingdom.

Vadim Ryabitsev

ONE SEASON IN THE TAIGA

1998. ISBN 0 9532990 0 7

Softback; sewn; viii + 179 pp.; 210 x 137 mm; 42 drawings, 4 maps.

Over many decades most Russian field biologists have looked forward during the long winter to their annual summer expedition. It is a chance to get away from the desk and laboratory work, and enjoy outdoor life for a few months, often camping in the pristine nature of some remote part of their enormous country.

Through his long career as an ornithologist, Vadim Ryabitsev's expeditions over many years took him to the far north, studying the tundra birds of the Yamal Peninsula. One year he opted instead for the taiga of the northern Ural Mountains. The objects of the study were arctic and willow warblers, but Vadim and his colleague also encountered many other birds, mammals and insects.

One Season is the charming account of that expedition, with its portrait of life in the taiga. Quite apart from its humour and first-hand descriptions of the wildlife, the book serves the serious purpose of showing how such an expedition is carried out and followed up, including details of how scientific observations are made. Vadim also explains why this kind of work is important. It is a magnificent tribute to the long-established Russian tradition of biological field work.

More recently this multi-talented author/artist has completed his field guide and hand-book *Birds of the Urals* (in Russian), illustrated with his own colour plates, and covering not only the Urals but also western Siberia. Copies are available from RNP.

'*Een uitstekend initiatief om Russisch werk in vertalig op de westeres markt te brengen!*' (Rob Bijlsma in *Limosa*)

'*One of the most enjoyable wildlife books I have read*' (Richard Porter in *RSPB Birds*)

Vladimir Smirin & Yuri Smirin

ANIMALS IN NATURE

1999. ISBN 0 9532990 1 5

Softback; sewn; x + 302 pp.; 210 x 138 mm; c.400 drawings, 10 maps.

Brothers Vladimir and Yuri Smirin are two more Russian biologist/artists. The need to observe behaviour of mammals in their natural surroundings took them on expeditions to many parts of the former Soviet Union - from Chukotka in the Arctic to the deserts of Central Asia, and from the forest near Moscow to the Sikhote-Alin Mountains on the Pacific coast. The various chapters in the book describe a wide variety of mammals and an equally wide range of habitats.

While *Animals in Nature* is predominantly of natural-history interest, wildlife artists will gain particular value from it. Chapter 1 'Sketching Animals' explains Vladimir's approach, and in particular points to some advantages that drawings have over photography in recording scientific details of animals' lives. Drawings can also be integrated with text far more closely than can photographic plates printed on special paper and separated from the text describing them. This principle has been respected in all of RNP's books so far, and in *Animals in Nature* there is a particularly close integration of text and illustration. The chapter-head drawings were made by Yuri especially for this edition.

Animals described include gerbils, jeyrans (gazelles) and other desert mammals; seals, walruses, sea otters, whales and arctic foxes; chamois, ibex and other Caucasus species; gorals; mole rats; elk (moose); pikas; burunduks (chipmunks); and mice.

'The line drawings [...] are absolutely delightful; and the text complements them perfectly [...] an ideal present for someone who loves both Russia and animals' (Helen O'Connor in *British East-West Journal*)

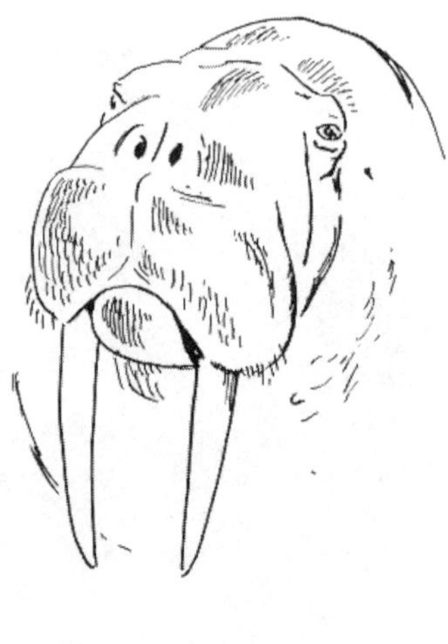

Feliks Shtilmark

HISTORY OF THE RUSSIAN ZAPOVEDNIKS 1895-1995

2002. ISBN 0 9532990 2 3

Softback; sewn; xii + 308 pp.; 210 x 137 mm; 30 drawings/maps, 28 portraits.

As governments and international organisations struggle with the gathering tide of environmental problems, a detailed knowledge of the biosphere - how it works, how it affects us, how we affect it - is becoming increasingly important.

The biosphere is so complex that our understanding of it is still no better than rudimentary. To increase our knowledge to an adequate level, we shall need examples of natural ecosystems that can maintain themselves on a self-sustaining basis indefinitely into the future.

This is the purpose of the Russian system of zapovedniks, or 'strict scientific nature reserves'. Conceived over a century ago in Russia, the system was built up in the Soviet Union, and the remaining Russian part of it has continued to expand in recent years. The zapovedniks comprise the only large-scale system of protected areas, with resident scientists, devoted primarily to increasing our understanding of the biosphere at local, regional and global scales.

Benefiting from his many years of experience working in zapovedniks, Dr Feliks Shtilmark has written the only comprehensive account of the system from its modest beginnings in the 1890s. The theory of *zapovednost* (essentially, non-intervention management combined with scientific research) is explained, and through a succession of historical periods we learn how closely the system has come to realising the vision. (Readers can keep up to date on recent developments by subscribing to the quarterly magazine *Russian Conservation News*; for details, see RNP website.)

'*Russian Nature Press [is] to be congratulated for the publication of this translation [...], which will bring Shtilmark's landmark study to the wider audience it deserves*' (David Moon in *Environment & History*)